FOREVER GREEN

FOREVER GREEN

The story of one of Canada's foremost foresters

Hector Allan Richmond

oolichan books

Lantzville, British Columbia

1983

Canadian Cataloguing in Publication Data

Richmond, Hector Allan, 1902-
    Forever green

  ISBN 0-88982-056-2

    1. Richmond, Hector Allan, 1902-
  2. Foresters - Canada - Biography.
  3. Entomologists - Canada - Biography.
  4. Forest management - British Columbia - History.
  I. Title.
SD129.R52A33 1983   634.9'092'4   C83-091479-X

Publication of this book has been financially assisted
by the Canada Council and by the Government of British
Columbia through the British Columbia Cultural Fund
and the British Columbia Lottery Fund.

Published by
OOLICHAN BOOKS
P.O. Box 10
Lantzville, British Columbia, Canada V0R 2H0

Printed in Canada by
MORRISS PRINTING COMPANY LTD.
Victoria, British Columbia

for Vi and Donnie

HECTOR A. RICHMOND

Hector Richmond's service to Canadian Forestry has now spanned almost half a century. He is amongst us as a living bridge to an earlier era when Public Service meant precisely what it states, when a forest entomologist appreciated the full range of interplay between man, trees and insects, and above all, when a leader led by inspiration, perspiration, imagination and good humour. Compared with the paler images that dominate today's forestry hierarchy, Hector is vital proof that outstanding individuals are still indispensable to the health of our over-organized society, that enthusiastic dedication to purpose is worthy, and that youth can be retained by shunning bitterness and cynicism.

For 25 years he served the Federal Government as a naturalist, forest entomologist, research leader and administrator. Besides inspiring and encouraging many young workers during the post-war years he enthusiastically supported and developed the National Forest Insect Survey.

He began serving the Forest Industry at an age when many are preoccupied with calculating their retirement credits. With his usual aplomb and versatility he became North America's first successful practicing consulting forest entomologist. He is one of three surviving charter members of the Association of British Columbia Professional Foresters, and a founder of the Western Forest Insect Work Conference.

In semi-retirement he continues to be a raconteur extraordinaire, and a keen amateur artist and actor. He has taken an active part in CIF for many years, never failing to participate in Section programs whenever called upon.

It is, therefore, with pleasure and pride, that we nominate

Hector A. Richmond

to the status of

Fellow of the Institute

The Board of Directors of the Canadian Institute of Forestry hereby grants Fellow status to Hector Richmond.

PREFACE

I am a professional forester, specifically a forest entomologist, and the era through which I have lived and worked, and of which I write, can best be described as the birth of a profession in British Columbia. It started at a period when technical forestry, and particularly forest entomology, was in its infancy, when there were few professionals in that science. It was also a period when professional work enjoyed a degree of independence and freedom not to be found in our society today. Government decrees, bureaucratic regulations, the dictates of labour unions and the regimentation of employee attitudes have since imposed restrictions.

My book has no morals to convey, no special message to impart, nor is it, in any sense, a book on science. It has given me, however, an opportunity to tell a few interesting tales of people and animals, and to express a few personal thoughts, convictions and speculations with respect to our environment and the land about us.

My life's work has extended through a period of more than fifty years of active participation in my profession, some thirty years with the federal government and twenty years as consultant to the forest industry of British Columbia. It has taken me through most of the forest regions of Canada in which I have worked, exclusive of Newfoundland and Nova Scotia. Through these years I have met many wilderness dwellers, living in almost total isolation, with their own philosophies and a mode of life adapted to their own particular circumstances.

Through the years, too, I acquired many friends in my profession, both in Canada and the United States. It has been my privilege to work with scientists foremost in their specialized field of work—men with international reputations. Some of these have contributed to this book through correspondence, some by reading and commenting on certain specific subjects contained herein, some by allowing me use of their personal research manuscripts.

For such assistance, I recognize, with warmest appreciation Dr. R. Blais, Research Scientist, Laurentian Forest Research Centre, Quebec; Dr. John Borden, Professor, Pestology Centre, Simon Fraser University, Burnaby, B.C.; Dr. N. E. Cerezke, Northern Forest Research Centre, Edmonton, Alberta; Dr. A. G. Davidson, Scientific Advisor, Forest Insect and Disease Centre, Ottawa; Mr. H. J. Irving, Director, Forest Protection, New Brunswick; Dr. John McLean, Professor Forest Entomology, Faculty of Forestry, University of British Columbia; Mr. Ross MacDonald, Director; Dr. George Puritch, Research Scientist; and Mr. R. L. Fiddick, Forest Insect and Disease Survey, Pacific Forest Research Centre, Victoria; Miss Shirley Mitchell, Pesticide Control Board, Victoria; Mr. J. A. Munro, Canadian Institute of Forestry, St. John's, Newfoundland; Mr. R. D. Peterson, Wild-rice Specialist, Department of Natural Resources and Environment, Manitoba; Mr. A. G. Racey, Secretary-Manager, Canadian Institute of Forestry, Ottawa; Mr. Don Robinson, Director, Fish and Wildlife Branch, Ministry of the Environment, Victoria; Dr. N. Schmitt, MD, FRCP (C), Victoria, B.C.; and Mr. W. J. Schouwenburg, Scientist, Canada Fisheries and Oceans, Vancouver, B.C.

I wish to thank the following for the use of certain photographs: the R. N. Atkinson Museum, Penticton, B.C.; Mr. Bill Taylor, Victoria; Jack Lindsay, professional photographer, North Vancouver; Mr. Alec Craigmyle, photographer, Canadian Forestry Service, Victoria; and Mr. Stan Dakin, Nanaimo, B.C., through whose skills and expertise pictures of unusual clarity and definition were reproduced from old faded negatives, considered by me to be almost useless.

For her care and enthusiasm in its typing, Linda Gauthier, Vancouver.

For her many hours of proofreading and her helpful suggestions and advice, my wife Vi, a vital partner in this book in both its writing and its action.

I am particularly grateful to the professionals of Oolichan Books, Ron Smith, Steve Guppy and Rhonda Bailey for technical assistance and direction in organizing and writing, and to Ruth McVeigh for her most constructive and helpful editing.

Finally I wish to thank an old friend and fellow forester, Mr. Doug McLeod, Forestry Supervisor, Western Forest Products, Vancouver, for his insistence through the years that I write this book.

CHAPTER ONE

The door slammed behind me. I called out, eager to tell Mother about the exciting job opportunity I had heard about.

Summer holidays had just begun and, although I could continue the after-school-hours and vacation job I already had in the entomology laboratory, I was more than keen to land this job with a real forest crew. Moreover it promised to pay double that which I had been making—the unbelievable sum of $100.00 per month. Not bad for a seventeen-year-old in 1919.

Word had come to me through the local forest ranger about a young forester who was in Vernon for the weekend. I learned he was in charge of a Forest Service cruising party whose cook had quit. They planned to leave the following Monday and as they were desperate for a replacement, I might be able to apply for the job.

Accordingly, I had called at the office of the District Forester, A. E. Parlow, where I met the man in charge of the cruising party. He looked me over carefully, questioned my cooking abilities, knowledge of the woods, experience in backpacking and so on. I assured him I was knowledgeable in the art of outdoor cooking and well qualified to do exactly what was needed. He advised me to come back on Monday morning when he might just possibly have a job for me on his cruising party.

Elated, I rushed home and told my mother what I had done. Since I was most definitely *not* qualified as a cook, being unable to prepare most of the things I had claimed I could, I saddled her with the chore of teaching me.

For the next three days all my time was spent making the sort of meals which would be expected of me on this work, leaving in my wake an endless array of spoilt food and dirty dishes. I gained culinary expertise by leaps and bounds.

Monday came. With my new-found knowledge (which, incidentally, stood me in good stead ever after) and overflowing with confidence, I beamed as I walked into the office of the District Forester bright and early. The verdict was simple and straightforward: "I have decided you are too small for the rugged work of backpacking necessary on this job. Even if you can cook, we can't use you."

That was it. I returned home very discouraged. My mother didn't seem a bit surprised.

Through the many years that followed, I saw a great deal of that same forester and although he became a very close friend of mine, I never let him forget his cold-blooded rejection of me as a cook! He was the late Chief Forester and Deputy Minister of Forests in British Columbia, Dr. C. D. Orchard.

Fortunately for my morale, I had a job to fall back on.

*　　*　　*

Two special friends of mine during my boyhood days were Bill Ruhmann and Douglas Gillespie. We were very closely associated in most of our early activities: wilderness wanderings, fishing trips, Boy Scout activities and in many aspects of nature study. In later years, the three of us attended Oregon State University and belonged to the same fraternity, Pi Kappa Phi. Bill and I studied forestry while Doug studied entomology in the Faculty of Agriculture. Shortly after our graduation, Doug died of heart trouble. Bill made a career for himself in logging engineering.

Bill's father, Max Ruhmann, and Mr. R. C. Trehearne were entomologists with the Provincial Department of Agriculture in Vernon. It was through my connection with Bill that I came to know them, and this had a pronounced influence on the future course of my life.

One day, when I was fifteen, Bill came to see me with some very exciting news from his father. Two jobs in the Entomological Branch were ours if we wanted them. The jobs entailed working two hours each day after school at a salary of $13 a month. During the summer holidays, we would work full-time at a salary of $50 a month, a fabulous offer for school lads. I hastened to my parents for their permission to take the job, a job which lasted for two years.

14

Because of this work and experience, I began to consider entomology as a possible career. I had always been interested in the outdoors and the forests. One of my principal hobbies as a boy was nature study. I had developed a collection of different bird's eggs, always being careful never to take more than one egg nor to disturb the nest. I started an insect collection, too. I even took a correspondence course in taxidermy but soon dropped it since it did not appeal to me.

Although seriously interested in entomology and also the forest as a possible career, I had little knowledge of what such a profession entailed, other than insects and trees. The life style it seemed to offer appealed to me as much as did the science itself.

Forest science and the attitude of the general public toward our forests and forest resources has changed greatly since my interest was first aroused in the science. I can well recall a prominent member of one of the larger forest companies, speaking to me in relation to forest insects and related problems. "I'm sure it's all very interesting," he remarked, "but we're really only interested in knocking the damn stuff down and moving on." He was not alone, however, in that attitude in those early days. The forests seemed inexhaustible, indestructible.

In such an atmosphere one might wonder how I could have become interested in a profession so little known or appreciated as forest entomology. The answer lay in the people I encountered in my after-school work in the Dominion Department of Agriculture entomological laboratory.

Through my association with the entomologist for whom I was working I was to meet Mr. Ralph Hopping, who had just been brought to British Columbia from California by the B.C. government. His work was to organize and direct a massive control programme against the Dendroctonus bark beetle in the ponderosa pine in the province, a programme similar to that which was currently underway in California and Oregon.

It was in this work that Ralph Hopping offered me my next job.

During the years 1920 through 1930, British Columbia was undergoing a devastating infestation of a bark beetle, known by the technical name of *Dendroctonus ponderosae (monticolae)* and

D. brevicomis. These two species were attacking and killing mature pine throughout most of its range in central B.C.

Bark beetles attack and kill trees by boring through the bark until the sapwood is reached. There, in that vital, life-giving region of the tree, between the inner bark and the wood, called the cambium, they feed, reproduce and raise a new generation. Upon reaching the sapwood the female projects a tunnel or gallery through the inner living bark, engraving slightly both wood and bark, along which she deposits her eggs. The resulting grub-like larvae, in turn, mine further through the cambium layer as they feed and grow, eventually transforming into young adults. When mature, they bore through the bark to the outside, fly away in search of a suitable host tree where the cycle is repeated. Trees are usually attacked simultaneously by thousands of beetles, and death of the tree follows within twelve months of being attacked. Since the larger, older or decadent trees are the preferred hosts, it is, to a great measure, nature's means of disposing of the older stands of timber to make way for the new, on-coming forests of tomorrow.

In the early stages of infestation, the British Columbia government undertook an extensive control programme hoping they might stem the spread of destruction wrought by this insect. Since the beetle spends its life under the bark, protected from any outside influence, no chemical application is practical as a means of control. The approved method of control at that time, was to fell the infested trees. Then they were limbed, barked, piled together and burned.

Several large camps of 30 to 40 men each were established throughout the area of control. Infested trees were identified and marked for cutting by a "spotter." It was as "spotter" I was employed, moving from camp to camp, marking trees and training new spotters how to recognize infested trees. Although I was engaged in the work for only two springs, the control work extended through most of the 1920's and into the spring of 1931, when it was terminated, having been recognized as an exercise in futility. The infestation and dying of timber spread at a rate far in excess of the control programme.

In due time, the beetle ran its course and disappeared after having eliminated practically all the old growth of ponderosa pine throughout the region. Today there stands in its place, young, vigorous pine, exactly the way nature planned it in the first place.

The bark beetle dictated the nature of my work and life over the next twenty years. During the second spring of the control programme, in 1924, I had my first face-to-face encounter with a bear. Because I was working alone, the Forest Ranger, Bob Little, loaned me a .45 Colt revolver with the comment, "It is a bad region for grizzlies, and in the spring, there's always the possibility that you may run into an old she-bear with cubs. You better carry a gun."

I didn't argue the point, and for two days I lugged the thing around despite finding it a nuisance and a hindrance. I had never fired a .45 Colt and had no desire to practice on an enraged grizzly. On the third day, having seen no bears, I left it in camp.

That morning, as I stepped from some rather dense brush on the summit of a small knoll, I startled a large bear eating something. He sprang around in a flash and, as he faced me, I saw to my surprise he had a completely white face. Later, when I recounted this incident to others considered knowledgeable on such matters, some diagnosed the bear as a "bald-faced grizzly," most uncommon and rarely seen. Others, equally qualified to express an opinion, insisted it could be only the reflection of my startled face in the bear's!

At the moment of confrontation, my one and only objective was to get away from the bear. Without a second thought, I leapt into the air, turned, and landed back in the brush, half expecting to hear the old bear after me. All was silent, however, so after a few seconds, I concluded there was no immediate concern since the bear was probably quite bewildered about what was going on.

I figured if I circled the base of the knoll I could approach the bear downwind and this might allow me to get close enough to have a good look without his knowledge. I started out and crept stealthily toward my quarry.

Meanwhile, back on the knoll, it must have occurred to the old bear that since I had appeared and vanished at one spot, that would be the place to avoid; making his safest course of escape down the opposite side.

The inevitable happened, of course: almost a head-on collision! We came face to face. Startled at seeing me so unexpectedly the bear slipped, skidded and almost fell over himself as he changed course to avoid me. I stood amazed as I watched him. Exerting every ounce of his strength, he brought his hind paws forward beyond his chin and when they made contact with the ground, his body catapulted through space. He raced across an open, grassy flat to safety down a ravine. Never before, nor since, have I had such a perfect view of a bear racing at full speed, as he thought, for his life. I was mighty glad I was not in front of him, racing for mine.

CHAPTER TWO

At the end of my second spring on control work and my first year of university, Ralph Hopping phoned and asked me to call at his office. He said he had an interesting job to offer me.

What he proposed exceeded anything I could have hoped for, and convinced me, beyond question, that forest entomology was the profession I wished to follow. Besides my interest in the work, the way of life it would provide held great appeal for me.

Forest insect investigations in British Columbia were initiated by the dominion government in 1923. The only previous work done in British Columbia had been bark beetle control under the B.C. Forest Service. With this change, Ralph Hopping now became employed by the dominion government in Ottawa. As there had been no previous investigational work or research in the problem of bark beetles in the province, it was decreed by Ottawa that a start should be made under his direction. The work proposed was a complete survey of the total insect population found within individual trees of pine and spruce in the region of the proposed study. It was planned I should spend the summer and autumn at the summit of the mountain bordering the east side of the Okanagan Valley, where the Kettle Valley Railway climbed over the mountain en route to the eastern part of the province. My instructions were to travel to Penticton via the paddlewheeler, the S.S. *Sicamous*. Once there, I was to open an account at a local store where I could order my food supplies and other needed items, then proceed by the Kettle Valley train to the famous "Lorna" trestle. I was to leave the train at a nearby water tower where it normally stopped to replenish its water supply after the long and arduous climb out of the valley from Penticton. I was then to find a suitable campsite, establish myself there for the summer and undertake certain specific duties as set forth in a work programme given me before leaving.

The equipment supplied me by the government consisted of

one tent, 12 x 14 feet, a canvas fly, sleeping bag, pots, pans, a coal oil lantern, a few tools, axe and a six-foot cross-cut saw. For the first ten days, I was accompanied by a fellow student, Norman Cutler, who later studied medicine at John Hopkins University and practised in New York City.

A suitable location for my camp was selected adjacent to the "Lorna" trestle, beside a reliable supply of fresh water. Our next task was to select trees known to be infested with bark beetles and wood-boring insects. These were felled, limbed and bucked into eight-foot logs. Separate cages were then constructed over trunk, limbs and stump. These cages were four feet wide, six feet high, and long enough to accommodate all the material that had to be caged. Since I had no lumber the framework was of poles and was covered with light factory cotton, several bolts of which I had brought with me. Once completed, it was necessary to examine each cage daily and to collect and preserve in alcohol all insect material that had emerged from the infested logs, limbs and stump. A total of six trees was cut and caged, fir, balsam, and two each of spruce and pine.

I had as my companion a small Airedale pup I called "Peter." With the departure of my companion Norman, Peter and I settled down for a four-month camp all by ourselves, the silence being broken each day only by a passenger train en route westward to Vancouver and one eastbound to the Kootenays.

Each day, the eastbound passenger train stopped at the water tower a mile along the track, where they would deposit any mail or express addressed to me. Although I rarely had an opportunity to speak to any of the train crew, they knew who I was and that I was alone. I made a special point of trying to be near the track as the train passed so I could wave to the engineer and let him know I was still around and that all was well. I later learned that the engineer always looked for me in case I needed help. The kindly nature of this engineer was revealed to me by a rather unusual incident.

Peter was only a small pup when I first arrived. His trip in the baggage car had left him with an intense fear and dislike of trains. I was walking down the track only a few days later when I heard the passenger train approaching. Between me and camp was the long horseshoe-shaped "Lorna" trestle, decked with galvanized tin for fire protection. Hearing the train, Peter decided the safest place for him would be under my bed in the tent at camp! He started to run. I ran after him, hoping to catch the pup before he got onto the trestle. The train was closing in on us and as Peter raced onto the trestle, I had no option but to step aside and let the train go by. There was the small pup, running his fastest, with a huge steam locomotive and a long line of passenger cars at his heels. The metal surface of the trestle added to Peter's difficulties as he slipped and slithered with each forward leap. The engineer, seeing the dog's predicament and my concern, waved assurance to me while he applied the brakes, slowing the train to a snail's pace. The train loafed its way across the trestle behind the panic-stricken pup. The dog won! Peter's pride in having outrun the CPR was no greater than my esteem for the big-hearted engineer.

Since the train normally passed my camp at full speed, it was necessary for me to mail my letters on the fly, in a manner commonly used on railways. A long willow stem was bent at one end so as to form a loop. The handle of this loop had to be long enough to reach the cab of the engineer. Letters for mailing were tied to this stick. As the train passed, the hoop was held in a position so that the engineer could slip his arm through it, remove the letters and toss the hoop back to me for future use. It was quite a tricky business.

The were other problems to solve. Having no stove, my only means of cooking was over a campfire and through a system developed the hard way. I eventually devised a method of baking a cake over the open fire.

On my first effort, I poured the cake batter into a tin and then pressed on a tight fitting friction top. I placed this in a larger tin and filled the space around the two with damp sand. Another tight fitting friction top was put on the larger tin. Finally, I rolled the whole thing in the coals over which my supper hung cooking in pots. Just as my meal was about ready, there was a terrific explosion. Pots, cake, kettle and fire went flying sky-high! I never did find the cake, although I recovered the empty pots. Obviously, there was something wrong with my design.

My later refinement, in which I made provision for the escape of steam, led to an outstanding success. By careful timing, I produced what was, in my estimation, a perfect cake. At least, it was edible.

I used much the same technology to cook grouse. After wrapping the grouse first in paper, then cloth, I tied the whole thing with string and placed it in a large tin of damp sand. Baking time was exactly 45 minutes and the bird was cooked to perfection. Grouse were plentiful in the area and with the aid of my .22 and this cooking technique I was able to enjoy one for dinner almost every day, from late August on.

While I was there, I had one of the most nerve-wracking experiences of my life, a situation created by my own thoughtlessness. A friend of my school days, Larry Lang, had come to visit and spend a couple of days with me. As the camp looked out on the long horseshoe-shaped "Lorna" trestle of the Kettle Valley Railway, we felt challenged to climb it. Together we discussed how best one might go about this feat and in the end we went down the ravine to the base of the trestle to assess the situation.

The trestle was made up of 12 tiers, each tier being approximately 15 feet in height and braced diagonally with timbers bolted to upright stanchions. We thought the trestle could be readily scaled by climbing upward between the diagonal bracings and upright timbers. I attempted a trial run and as it seemed easy enough, continued on. As always, however, things do not look as high when viewed from the bottom as when seen from the top. After scaling the trestle for about half the distance, I stopped to rest, and for the first time, looked down. My hair stood on end. I looked up at the amount of climbing ahead of me. At that moment I figured the better part of valour was to give up the idea entirely and go back down. Here, however, a problem arose which I had not previously considered. Going down is a very different thing from going up; in fact, it soon became evident that since I could not descend with any degree of safety, my safest course would be to continue upward and to crawl onto the top of the trestle. With increasing weariness and decreasing energy, I eventually reached the top. Now I was faced with an even greater and unexpected difficulty. The deck of the trestle extended about three feet in width beyond the upright stanchions on which I was climbing. I had, therefore, a ceiling above my head. To get to the top, it would be necessary for me to go out and around the three-foot overhang. I was panic-stricken. Again, I considered going back down, but abandoned the idea as being too dangerous. I would have to get out and around this overhang some way.

The top of the trestle was made of squared timbers running crosswise in the manner of railway ties and spaced about six inches apart. Along the outside perimeter was a margin of squared timbers bolted to these cross-members making a rim for the full length of the trestle. If I could just reach out and get a good grip on one of these fringe timbers, I reasoned, I might be able to pull my body out and over the edge and on to the top. After much thought, I decided to try, since I could think of no alternative.

I wrapped my legs firmly around an upright stanchion, reached up with my hands, grasped the cross-ties and leaned backwards. I was thus able to reach the outside edge of the overhang. At this point, I let go with my legs, gave a good kick, swung my body outward and with a heave, some twisting and squirming, finally got my chest over the edge, followed quickly by my body and legs; so at last I was lying on top of the trestle. I was a nervous wreck, too shaken to stand, scarcely able to speak. I lay there in a state of shock while I regained my composure.

My friend Larry, who had been watching in great concern, came hurrying to me to express his relief in seeing a safe conclusion to this pointless, danger filled adventure.

CHAPTER THREE

I was awakened at an early hour by Peter's defiant barking and an accompanying voice. "Call off your dog. We're the police." I sat up in my sleeping bag and looked out the tent door. Two uniformed men stood there waiting.

"Peter," I called, and the dog returned to me, reluctantly substituting his role of vigilance for that of a welcoming committee. "Come on in while I put on some clothes." I was naturally very curious as to why two police officers should appear at such an unearthly hour and in so remote a spot.

"Are you alone?" was their first question.

"Yes, except for my dog. I expect to be here until autumn."

"We heard you were here so we came up from Penticton on a freight last night. We'd like to stay in your camp for two or three days. We're looking for a man who murdered another man yesterday." They then told the story. There was a railway work train parked on a side track some three miles beyond my camp. A dispute between the foreman and one of the workmen had ended with the foreman being stabbed to death and the culprit swearing similar vengeance on one of the others before he disappeared into the surrounding woods.

"We figure," they continued, "that the man might return to the train at night to cause further trouble or to steal food. He might even show up here."

If there was a murderer lurking in the woods and any possibility of a visit from him, I was more than happy to have the police in my camp. "You're certainly welcome here," I said. "I've got plenty of food. Right now I'm going to get a fire going and cook some breakfast for the three of us."

For the next three days, the police officers lived with me, alternating their time between my camp and the work train. Finding no further evidence of the culprit in the region, they assumed he had slipped away. Whether or not he was ever arrested, I never heard.

That, however, was not the only murder to add variety to my life that summer.

A Forest Service telephone line went past my camp and extended to a fire lookout tower at the top of Little White Mountain. Draped from tree to tree, the line had been out of commission during the summer, and a repair man, Jack Dunn, had been sent to work on it. It had been arranged that he would live at my camp. So, for the time he was there, I had the luxury of a portable telephone, which had been attached to the wire.

Early one morning, while I was still in bed, the phone rang. Under such circumstances it suggested something of unusual importance.

"This is the Kelowna police," the voice said. "The Forest Service tells me that you are camped beside the 'Lorna' trestle."

"That's right."

"Well, there's a fellow headed your way who murdered a man in Kelowna last night. We think he has taken a trail running up the mountain from Kelowna to the Kettle Valley Railway where he can jump a freight when it stops for water at the water tower near you. Keep your eyes open for him. If he shows up, phone us."

This was a most unexpected bit of excitement for Jack the lineman and me. Together we discussed the matter in some detail. The trail from Kelowna extending up the mountain, reached the railway some two miles east of my camp. This was the trail on which it was assumed he would be travelling. It was evident from our experience that if he walked all night he might be expected in late morning.

The section gang who kept the line in repair lived in a railway section house located five miles west. The foreman was a man by the name of Joe Martino, who operated a gas speeder. We reasoned that if Joe took his speeder, he should be able to find the alleged murderer somewhere along the rails.

About fifteen miles beyond the section house was a trail running down the mountain to the valley below. The trail would be the easiest and shortest route to reach Penticton where the fugitive would have ample opportunity to jump a freight to Vancouver. If Joe could pick this fellow up on his speeder and induce him to take this trail rather than stay on the tracks, it would be a simple matter for the police to apprehend him at the other end of the trail.

With this in mind, we contacted the police. They accordingly got in touch with Joe Martino who agreed to follow the suggestion.

The suspect, apparently seeing my camp, skirted around it so we never saw him. Further along, however, he was spotted by the section boss, given a ride and dispatched down the trail as planned. The police were then notified and were waiting. He was subsequently taken into custody.

His trial was held in Vernon at the fall assizes. Out of curiousity I attended. It was alleged the murder had been committed during a period of extreme drunkenness, and that on awakening in the morning, the accused had no recollection of stabbing the victim. A record of the liquor he had bought was exhibited, since at that time a permit was required for any liquor purchase. He claimed that as he wished to go to Vancouver the next morning, he had walked up the mountain by trail hoping to board a freight at some sidetrack or water tower. This story was accepted by the court as being reasonable, and although he was found guilty, a moderately light sentence was imposed.

From my experience, this was an impossible story. In no way could he have covered the ground he did in the time he claimed, particularly after such a drunken night. It seemed obvious to me he had travelled through the night. I discussed this with the police after the trial. They felt the killing was actually a retributive affair within the Hindu community and through it, in the opinion of the accused, justice had been done.

The two summers that I spent in that area were both enjoyable and memorable. The individualism that characterized the Entomological Branch in those early days has long since disappeared in today's ever-increasing bureaucratic expansion of government departments. I really loved the type of life the work offered. Despite my isolation, I seldom experienced the slightest feeling of loneliness. The mere fact that I could see and wave to the passing passenger train each day was enough. And, of course, I had my dog, Peter.

CHAPTER FOUR

After two seasons in this location, in 1927 the nature of my work changed. I was sent to the forest regions of Alberta where I experienced several of the most interesting and adventurous years of my working career. I was then in my third year of university.

The devastation caused by bark beetles in the forests of British Columbia aroused the concern of the forest officials in Alberta to the possibility of a similar situation developing in that province. Accordingly, they requested someone be sent to undertake a survey of the forest insect conditions within the Province of Alberta, which were, at that time, under the jurisdiction of the Dominion Forest Service, Ottawa. To this work I was assigned.

For the next three summers, I travelled mostly by saddle and pack horse on the eastern slope of the Rockies covering the forest regions of Alberta from the International boundary, north to the Athabasca River, including all of Banff National Park. At the same time, my work included the Cypress Hills of southeastern Alberta and southwestern Saskatchewan. This is rolling, hilly country, covered predominantly by poplar and lodgepole pine. While I was in that region I seemed to be plagued with bad luck, for myself and anyone who happened to be with me.

After spending several days in the Cypress Hills with the Forest Supervisor, Mr. Parker, I planned to leave by train from Maple Creek. Fate, however, decreed otherwise.

Late on the final day while en route to the Forest Service headquarters, I got out of the Model T Ford in which we were travelling to open the range gate. I noticed oil dripping from the motor. On closer examination, I discovered the screw plug on the oil pan was missing and with it most of the oil. There was no alternative but to walk three miles to the nearest ranch, phone the Forest Service and have someone bring out another plug and some more oil. Having done this we had the good fortune of being invited to eat our supper at the ranch. With the arrival of assistance, we again became mobile and eventually returned to headquarters, completing our day at about midnight.

Early next morning, we departed for Maple Creek where I was to board the westbound CPR. After three miles, the front spring of the Model T collapsed. The only solution for this was to return, dismantle the front spring from another Model T and replace our broken one. By the time the repair was completed, our race for the train became critical. We could still make it, barring further delays. Again we started for Maple Creek. Halfway there, one of the tires blew, necessitating another delay while the spare replaced the flat. By now it was just a question of how fast the Model T could be induced to travel. Then, only three miles from the station, one of the rear tires went. Without another spare, we were stranded, while the westbound passenger train arrived and departed. Eventually we dragged into Maple Creek; I procured a room at the hotel for the night and my host returned to his home in the Cypress Hills with one final warning: "When you get on that train tomorrow stay put or you'll probably be plagued with more bad luck."

The next day I boarded the westbound train, still remembering the warning I had received. My berth on the train was an upper standard. Upper berths were separated one from the other with a wooden partition that slid into place when the berth was made up.

I had climbed the ladder into my berth and undressed, when I suddenly realized my pyjamas were in my suitcase, which was on the floor under the lower berth. To get the pyjamas meant dressing again, climbing down the ladder, getting them out of my suitcase and climbing back to my berth. This seemed quite pointless. The simple thing was to forget the pyjamas and go to bed raw. After all, who would see me anyway? I got into bed naked and was soon asleep.

All went well until about four in the morning when I was awakened with a terrific jolt, followed by a stunning blow on my head. By the time I was fully awake and conscious, I found to my utter amazement that I was no longer in my own bed. I was in the next berth forward and what is more, I was not alone! I was in a girl's berth, my naked body, only partly swathed in bedding, lying on top of hers.

Apparently, I had shot forward as the partition between the two berths slipped out and had tobogganed into her berth, receiving a blow on my head en route as I collided with the wooden partition. I never did find out just what she thought of this sudden turn of events. We didn't discuss it, nor did we introduce ourselves! I simply wiggled back into my own berth, taking my bedding with me.

Two hours later I opened my eyes to become aware of the fact that it was morning and I should be up. I began to recall the events of the night, the jolt, the blow on my head, my visit in the next berth. Then I realized the train was not moving. Completely bewildered, I dressed and climbed out of my berth only to find the car cold, totally empty of all passengers, the windows broken, the mirrors in the men's dressing room smashed, chairs upside down, everything in a shambles and covered with dust. Mystified, I stepped from the train to be greeted by passengers pacing up and down the track. Ahead lay a mountainous pile of wreckage; two locomotives mushroomed into each other with the baggage cars of our train piled on top. Trapped within the wreckage were the engineer and fireman of our train, killed at the instant of collision between our passenger train and the eastbound freight.

My first question to a fellow passenger was, "What has happened?" as if it was not perfectly obvious. "Where the hell have you been?" was his reply. He told me they had been pacing around for the past two hours to keep warm. He said that a doctor and nurse had been checking each passenger for possible injuries and suggested I search them out and report my experience to them. Apart from a severe headache and a very stiff neck, my condition seemed to be fair enough, so I ignored the suggestion, a foolish thing to do, under the circumstances. It did seem rather odd to me that no one had checked through the train for any passengers still in bed, as I had been.

The tragic night had one bright spot. A relief train arrived and all passengers were taken on board and served a free breakfast, compliments CPR.

I never did find the girl I had so unceremoniously visited during the night.

CHAPTER FIVE

When I was 12 years old, my sister gave me a diary, and from that time on, I have kept a daily record of events. It is mainly from these I have drawn the account of my life for without them, much would have been forgotten.

I was born on December 21, 1902 on a cattle ranch on the Cimarron River in Oklahoma. My father, Thomas Richmond, was born in 1866 in Scotland on a 2,000-acre family estate named "Balhaldies," located in the vicinity of Blackford and Gleneagles. There he lived until the age of 17, receiving his education at Kings College, Chester, England and in Switzerland. An estate such as "Balhaldies" operated on a near feudal system, the owner reaping the profits from the work of others who lived and worked there.

The estate was made up of several individual units, each with its own name. With my wife Vi, I visited it in 1977 and although it was sold in the late 1800's, it is still known as "Balhaldies." It is now broken into several individual farms, each bearing the name of the unit by which it was called when part of the original estate. The entire property has been little altered over the years, in marked contrast to the speed with which rural land is subdivided, "developed" and, in many cases, decimated on the North American continent.

My father's diaries and records show the difficulties of the times. The estate was in grave financial trouble and he could see only a bleak future ahead. Accordingly, at the age of 17, he decided to seek his fortune in the wild west of America. In 1884, he left his Scottish home and imigrated to what was then the territory of Oklahoma.

The early American West was developed largely by Scottish cattle companies, according to the curator of the Museum of the Early West, in Trinidad, Colorado. One of the first of the big oufits was the Muskateen Cattle Company, with headquarters in Edinburgh, Scotland. My father had been hired to ride with this company. For the privilege and experience of learning to be a "cowboy," he was required to pay 500 pounds before leaving Scotland. I never learned how long he had to ride before qualifying as a wage earner, but I know he rode for the Muskateen Company for three years, after which he decided to start a ranch of his own. He located on the Cimarron River, on the Panhandle of Oklahoma.

In 1892 while visiting his sister's family in Emmitsburg, Iowa, an area to which many people from the Old Country were emigrating, he met Alice Neale, a young girl whose family had recently arrived from Western Supermare, England. A year later, they were married. He took her to his ranch in Oklahoma, where between the years of 1893 and 1907 they raised a family of five children: Alice, Donella, Gwendoline, Thomas and myself.

The Panhandle of Oklahoma was a very remote region and since there were no schools, the neighbouring Clarks together with the Richmonds erected a small school house and hired a teacher. There the children of the two families received their early schooling—although I was not school age at the time.

My father, a pioneer of the early west, rode and lived the life of the traditional cowboy and eventually owned and operated a large cattle ranch, yet he remains in my memory as the true Scottish gentleman he was. Never once in my life did I ever hear him swear or use a profane word. Sundays were always rigidly observed as a day of rest when all ranch work ceased except care of the livestock and jobs demanding immediate attention. We all wore our best clothes throughout the day.

My recollections of our life in Oklahoma are rather vague, as my father sold the ranch before I had reached school age. We then moved to Oregon, where he bought a farm near Dayton, in the Willamette Valley. After three years, he again sold and we moved to Vernon, British Columbia in 1911.

When I was a boy, Vernon was a small town with a very pronounced English atmosphere. Schooling, the climate and the British flag were the compelling reasons for our move, although my father never ceased to praise Oklahoma, the land that had given him so much in his early years. For me, it was a very happy move, as Vernon had so much to offer a boy. My father bought a farm close to schools, but adjoined by miles of woods where one could explore and walk for hours, never seeing a soul. I spent many days in these woods. I even built a log cabin and installed in it a cook stove made from a four-gallon oil can. I cooked meals in my wilderness retreat, and, amazingly, never burned the place down, although more than once I was forced outside to escape the searing heat when the stove got white hot from burning an overload of resinous wood.

My boyhood days were probably typical of the average boy on a farm at that time. Morning and evening chores throughout the year included the feeding, watering and care of all the livestock: cattle, horses, cows, pigs and poultry. As well, I had to take care of the daily wood supply for the house. Summers were taken up largely with putting up the hay crops. The alfalfa was ready for cutting as soon as school was out for the summer. This was followed immediately by cutting the timothy hay. By the time that was completed, the second cutting of alfalfa was ready. By the end of summer a crop of clover was cut off the timothy fields. Between hayings there were numerous other farm projects, particularly in the orchard. During the school year, most of my Saturdays were taken up hauling hay from the stacks in the fields.

As far back as I can remember, I always raised a pig of my own. It would be assigned to me in the spring and sold in the fall along with the others. On one occasion I entered a pig raising contest sponsored by the provincial government. My pig won first prize, $7.00 as I recall, as the best pig in the local club. During the second year, my next pig won the local competition and also the provincial prize as the best pig in the province. My provincial prize was a purebred sow, shipped to me from Crossland Brothers in Duncan. She was registered under the name of "Marie-of-Cedar-Creek." Shortly after I received her, she surprised me with a litter of 13 piglets. Because they were purebred, it was desirable to register them and to receive official papers of their heritage. To do so, I had first to become a member of the Swine Breeders' Association of B.C. This completed, my next task was to select names for each pig, which became quite a game for all of us kids. One name I selected stands foremost in my mind, that of a teacher I particularly disliked. She was incompetent, her disposition particularly miserable, and as she came near, she left an aura of stale perspiration in her wake. In recognition, I named one of my pigs for her. Her name thus became immortalized in the official records of the Swine Breeders' Association of British Columbia. Naturally, I never advised her of the honour I had conferred upon her.

Of all my youthful activities, however, nothing contributed more than the years I spent in the Boy Scouts under my Scout Master, the late Charles F. Morrow (who later became a judge). He developed in the boys a degree of discipline and interest only a man of his dedication and calibre could achieve. One of the most memorable events of our Scout troop was the annual production of a three-act play. Produced exclusively by the boys, it ran for three nights in the Anglican Parish Hall. From this we made enough money to operate our troop for the year and to finance a two-week summer camp on Okanagan Lake. It was there I had my first experience in acting, something that has always had great appeal to me. Although acting never developed as anything more than a hobby, there was a time when I might have considered it as a career, had the opportunity presented itself.

My brother Tommy, four years my senior, was an inventive genius. One of his most memorable contraptions was a toy fire engine which afforded us many hours of enjoyment. The truck itself measured about two feet in length. Mounted on top of it was a pint-sized tin with a tight fitting friction top and a second smaller tin also with a friction top. These two tins were connected by a small metal tube. The larger tin served as a water can and the smaller tin was filled with vinegar. A length of fine rubber tubing led away from the water can and terminated in a small metal nozzle which my brother had also made. To generate pressure in the water tank, a spoonful of soda was dropped into the vinegar and the lid quickly replaced. The resulting effervescence produced an appreciable pressure within the water tank and a steady stream of water was emitted through the nozzle. Obviously, this was an outdoor toy!

Its efficiency as a fire engine was tested by burning houses which we made from cardboard boxes. In preparation for the fire, a string was stretched so that it would burn through as soon as the model house was aflame. With the burning of the string, a cowbell would be released and at the sound of the bell, the fire engine would rush to the threatened building. To make it more realistic, we arranged it so we could not see the start of the fire, nor realize how bad it was until we were called into action by the ringing of the bell. The game became very challenging and we elaborated on it as time went on. We eventually developed two hose lines leading from the water tank with an increase in the size of the vinegar can.

All went well with the fire engine until a real obstacle arose—Mother. When she discovered that her vinegar supply disappeared almost as soon as she acquired it, she put a stop to its further use. Tommy was ingenious, however, and was not to be deterred by such a small matter. Rhubarb juice was substituted for vinegar. Since my father grew large quantities of rhubarb commercially, it was readily available, and the juice was easily extracted by putting the stalks through a cider press. But my father put an abrupt conclusion to this practice when he noticed the rhubarb being sadly depleted just when he had a ready market for all we could grow. That ended our fire engine, and perhaps it was just as well for we had pretty well exhausted the novelty and challenge of fire fighting.

In the First World War, Tommy joined the Royal Canadian Engineers, went overseas and was killed in France a month before the end of the war. He was only twenty.

Unforgettable too, is the day I unintentionally rocked the Vernon high school with an explosion!

Another chap and I found the formula in an old chemistry book. Prepared as directed, it would produce a substance requiring only agitation to set it off. All one had to do was to combine certain chemicals with a liquid, allow fifteen minutes for the chemical reaction to take place, drain off the liquid, allow the residue to dry and the result was a very explosive substance.

We prepared the concoction during the noon hour, spread the mud-like residue over a sheet of paper and placed it in an empty desk at the back of the school room. During the latter part of the afternoon, we agreed one of the boys would take it from the desk and place it over the hot air register in the floor. There it would finish drying and, we hoped, explode, startling all the girls as well as our teacher who was a very meek and ineffectual individual. We reasoned that with the small amount we planned to prepare, the explosion should be no more than a fire cracker.

By the time the paper was removed from the desk, it had already dried. As the boy picked it up, the paper bent and the fringe edges of the smear began to crackle and pop. Frightened, he dropped the sheet. The paper turned over and floated gently to the floor, coming to rest on the hot air register, smear side down, so that it looked like nothing more than a sheet of white paper. It continued to crackle and pop in reaction to the hot air. The teacher, hearing this strange sound, came slowly down the aisle to where the sound originated and asked, "Who's striking matches back here?" There was no answer. We all waited in tremulous anxiety as to what might happen next. Noticing the paper from which the strange sound emanated, he gave it a kick. It exploded! A white flash shot up about six feet into the air, accompanied by a bang fully as loud as a twelve-gauge shotgun. The poor teacher was practically paralyzed with shock. So were we. He stood motionless and speechless as the other teachers rushed into the room. There were only four high school teachers in Vernon at that time. The principal, bewildered as the others, asked who was responsible. As all the boys had been involved to some degree, each boy put up his hand. After a moment's hesitation, he asked to see us after school.

His first question was, naturally, just what had we done to create such an explosion. We showed him the book and the directions we had followed. He then warned us most emphatically of the dangers inherent in playing with unknowns and how such things could lead to tragic consequences. We all left feeling very sheepish, decidedly scared, but thankful the adventure had led to nothing more than a loud noise, a shocked teacher and an impressive lesson for us. One thing was evident—the principal, Tom Calder, really knew how to handle boys.

*　　*　　*

During the last year of the war, food production became a critical issue. To assist in the work on farms, a movement, known as "Soldiers of the Soil" (S.O.S.), was launched by the provincial government. By this plan, any boy between the ages of 15 and 19 whose school principal could recommend him for passing without examinations, was excused from school in early spring to do farm work as a Soldier of the Soil.

To me, it was a real wartime bonus, and I wasted no time in enlisting. While my father was none too pleased to see me lose some six weeks of schooling, he was, nevertheless, desperate for farm help. With my bronze medal, on which was inscribed the letters S.O.S., I left my classroom May 15th, with a distinct feeling of superiority in being a boy with a farm to go home to, leaving those less privileged to sit through another six weeks of schooling.

A total of 1,130 boys were enlisted and placed on farms in the province, with another 100 sent to farms on the prairies. Wages ranged from $5.00 to $75.00 per month, plus board. I don't remember how much my father paid me, but I received free feed for my pigs, which in itself was of considerable worth.

For the purpose of supervision of working and living conditions, British Columbia was divided into seven zones, each with a volunteer superintendent and local supervisory committee.

That autumn, there was a severe outbreak of influenza. Thousands of people were seriously ill and many died as a result. Schools were closed and public gatherings prohibited between October 21st and December 2nd. As a year of schooling, 1918 was a disaster; from my viewpoint, as a kid, it was a fine year.

My boyhood days in Vernon were wonderful and the Okanagan Valley was a beautiful place in which to live, almost untouched by the march of progress and the influx of its present population. As I look at the valley now and compare it with those bygone days, I see something lost, never again to be the same. Industrial developments, housing projects and highways have replaced productive orchards and prime agricultural land in many parts of the valley.

The multicoloured waters of Kalamalka Lake have lost their brilliance and distinctive hues. Some sand beaches are now spoiled with the accumulation of silt and mud. Marshes on Kalamalka Lake which once filtered residues from the inflowing water have been channelled and drained. Many of the shoreline waters of the lakes throughout the valley are plugged and tangled with Eurasian Milfoil weed. The control and elimination of

Eurasian Milfoil has long been recognized as impossible wherever it occurs elsewhere on the continent.

Waters have become polluted with accumulations of sewage and effluent from surrounding communities and this has, through the years, contributed greatly to the establishment and flourishing of the Milfoil weed. The runoff of chemical fertilizers, herbicides and insecticides has changed the chemical composition of the lake to such an extent that the provincial government in 1979 warned the public against the consumption of fish taken from Kalamalka Lake. Then, as if man had not already done enough damage, the army disposed of unused ammunition and bombs by dumping them there at the end of the Second World War. They have now drifted shoreward, embedding themselves in the sand and silt and constituting an ever-present menace to any attempt to remove the Milfoil by mechanical means. This is one of the strongest arguments used for the continued use of 2, 4-D in Kalamalka Lake.

On the brighter side of the picture, the City of Vernon has developed a noteworthy system of sewage disposal so that, instead of increasing pollution, the sewage is now purified and the water recycled for use in the irrigation of thousands of acres of otherwise non-productive agricultural land. The problem now facing the Vernon disposal system is, apparently, the need for more readily accessible land to absorb all the recycled water, a situation accentuated by the recent wet weather pattern.

Although some of the damage done to the valley is irreversible, it remains a place of charm and beauty. But I'm glad I knew it in an unspoiled state and can remember the valley as it was when I was a boy. It will never be quite like that again.

CHAPTER SIX

I completed my four years of undergraduate study in forestry at Oregon State University, Corvallis, Oregon. My life as a student was not greatly different from that of any other self-supporting scholar. I lived my first winter in a small room attached to a woodshed, which I and my companion, Doug Gillespie, rented for $3.00 per month. I took a job in a campus restaurant as second cook, working from 6:30-8:00 a.m., 12:00-1:00 p.m. and 4:30-7:00 p.m., for which I received my meals. After my first winter, I became affiliated with a fraternity, Pi Kappa Phi, and thereafter moved from my woodshed residence to the fraternity house.

With the money made during my summer's work, together with some borrowed from the Student Loan Fund, some from my sister and some I earned through a fairly profitable sign-painting business I developed as a sideline, I managed to finance my undergraduate education. In addition to earning or borrowing, however, there are other ways of cutting expenses.

At the time I attended Oregon State University (1924), there was an out-of-state fee of $50.00 a term ($150 per year), which was a lot of money in those days. There was also two years' compulsory military training for all American students. This "war" as it was called, was regarded by many, including me, not only as a nuisance but a waste of four precious hours a week. It was a "must" for all, except foreign students. It was obvious to me from the outset that I had two obstacles to surmount: to obtain exemption from the out-of-state fee and to become exempt from military training as a foreign student.

During my first two terms, I paid my out-of-state fee and registered in the cavalry division of the R.O.T.C. I figured I could at least learn what the Army had to teach about horses.

Next I applied for exemption from the out-of-state fee on the grounds that my family had once owned a farm in the Willamette Valley, Oregon, during which time my father had paid Oregon a lot of taxes. I was, furthermore, registered in the military as required of all residents. On the basis of this and because I had no money with which to pay the fee, the Board exempted me from further out-of-state fees.

I then requested exemption from the military. Since I was a Canadian, with receipts for having paid my out-of-state fee for the term, my request was granted without question. I really didn't feel too badly about this, for we had previously been taxpayers in the State and I was certainly no loss to the U.S. Army.

Oregon State's R.O.T.C., had for many years enjoyed the prestige of displaying a gold star on the sleeve of each man's uniform, a recognition of excellence awarded by a Military Review Board from Washington, D.C., at their annual inspection. This year, the University lost the gold star. To be more exact, the cavalry section lost it for them, and I, of course, was in the cavalry.

The day of the annual inspection, we were all assembled at the stables in preparation for our display, only to be advised that the Inspection Board had laid down a few requirements with which we were not familiar. First of all, we were to use bridles with both curb and snaffle bits, and two pairs of reins. We had never ridden this way before and controlling a horse under such a condition was foreign to all of us. Secondly, we were to use sabres in the march-past. Each would draw his sabre, salute as he passed the review stand, after which he would return his sabre to its sheath. This also was a "first" to us. We were carefully drilled in the correct procedure and warned that because of the slight curvature of the blade, if the sabre was returned to the scabbard the wrong way around it would stick and would be very difficult to withdraw. After some drilling we seemed to have the march-past and salute pretty well perfected.

Prepared for the rigid inspection, we moved on to the parade grounds, a large field on the outskirts of town. Our first exercise was to be a ride past the inspectors, single file, with sabres drawn, saluting as rehearsed. The review officers asked for a repetition, probably because the performance was unbelievably poor. Unfortunately about 50 percent had returned their sabres backward into the sheath. In the second march-past, therefore, the only salute reviewers received from about half the company was the rear ends of soldiers hunched over their saddles, holding the scabbard with one hand while pulling on the sabre with the other.

After this fiasco, we were lined up, bawled out for having created such a schmozzle and advised our next display would be a charge. Starting from the far end of the field, we were to bring our horses to a stop in a straight line abreast the review stand.

"You fellows follow me," the sergeant ordered. "When I stop, you will pull up abreast of me and come to an orderly halt."

When all was in readiness, horses in an even line in extended order, the command came: "CHARGE!"

Unfortunately, the horses did not seem to understand the difference between "charge" and "race," and we did not realize our inability to control a horse using these fancy bridles with the curb and snaffle bits. In any case my horse was bound and determined none would pass him, and furthermore, he felt he could make much better time if he jumped over each bush and weed. Away we went at breakneck speed, down the field, over the bushes, past the sergeant who had come to a stop as planned, on past the review officers to the far end of the field, coming to a dramatic finish when the entire company piled into the fence. Wire and rusty staples shrieked and squeaked as the horses milled around tangled in the wrecked fence, while the poor sergeant, who we had abandoned in mid-field, came riding down, embarrassed, humiliated and fuming. He was an ex-British Army Cavalry Officer, and his command of profanity was an art. For the next few minutes the air was blue with what he called us and his threats. "We're going to do this charge again," he barked, "and this time it will be done exactly as ordered." Again he would lead us. Again he would stop in mid-field before the review stand. We would come abreast of him and halt in an orderly line.

Untangled from the wire, we took out positions and with the order "CHARGE," we were off. As planned, the sergeant led, and with his hand signal, ordered a halt of the charging troops behind him. Once more the roaring cavalcade charged down the field, rushed past the sergeant and the amazed Washington Brass, my horse still leaping over every weed he saw as if it was a hurdle, and so we continued, coming to a final halt only when we all piled up in the fence at the other end of the field.

This exercise now completed, the reviewers were presumably satisfied that nothing further could be gained from our charge tactics. We were then paraded in a long line, standing in front of our horses with rifles at "present arms" as the inspectors strolled past in a very dignified procession. Pausing, an officer seized the rifle from the chap standing next to me. In a very officious and brisk manner, he tossed the rifle back to the student with the obvious expectation he would seize it and just as smartly return to the "present arms" position. The student, taken by surprise, threw up his arms as though he feared being hit in the face. The rifle hit his funny bone; the student shouted "ouch" and the rifle clattered to the ground.

A few minutes later a final episode occurred. We were again mounted on our horses at rigid attention before the review board. One of the horses figured he had had enough of this horseplay and broke ranks, headed straight for the inspectors, forcing them to scatter, and rearing and bucking sent the rider flying through the air to land flat on his stomach at the feet of the astounded officers.

This concluded the cavalry division's spectacle before the reviewers from Washington, D.C. The debacle was followed by the loss of the gold star for the year to Oregon State's R.O.T.C. To the military school it was a disastrous day. To me, it seemed very funny.

Although the great depression of the 1930's was not yet upon us, jobs were already becoming very scarce. Unlike most of my classmates at the time of graduation, I already had a job to go home to. Mine was a continuation of the work in which I had been engaged during the previous summers in British Columbia and Alberta.

The total staff responsible for all the forest insect problems in the two provinces numbered four: Ralph Hopping, George Hopping, Bill Mathers and myself. The headquarters for this work was Vernon, B.C.

A year later, June 3, 1929, the most important event of my life occurred. I married Vi.

Vi was born in New Brunswick of English parentage. Her

father served with the Hudson's Bay Company through most of his life, eventually becoming manager of the Victoria and Vernon stores. Vi completed her elementary schooling at St. Margaret's School for girls in Victoria and St. Michael's School for girls in Vernon. Her high schooling was completed in Vernon. Although I was several years ahead of her in years and in schooling, I became acquainted with her about the time she finished high school and dated her as often as opportunity permitted. Such occasions were infrequent since my time in Vernon was restricted to brief periods between my return from university and my departure for summer work in the woods and Christmas vacations. Added to my difficulties was the fact that Vi was *the* "glamour girl" and I was but one of many interested in her. This struggle for supremacy continued through my university years. Just as Vi was preparing to depart from Vernon to enter the nurses training school at the Royal Victoria Hospital in Montreal, I shattered all such plans by marrying her.

The wedding took place in Vernon, a full choral service at the Anglican church. The service was performed by Bishop Doul, who had known Vi since she was a little girl. From then on, she was destined to live a rather nomadic life in the woods and the wilderness regions of Canada and to reside in such places as Vernon, Winnipeg, Ste. Anne de Bellevue, Quebec City, Victoria and eventually Nanaimo on Vancouver Island.

Although raised as a city girl, Vi's heart was always in the country and she loved nature. She was an ideal partner for the kind of life we shared. From our wedding day forward, she became an integral part of my life and my story.

Following the wedding, Vi and I enjoyed a brief honeymoon on the B.C. coast, visiting Vancouver, Pender Harbour, and Princess Louisa Inlet, which in 1929 was little known to anyone. A beautiful wilderness lodge had been constructed at the extreme head of the inlet by a Mr. MacDonald. Unfortunately it burned to the ground shortly after our visit.

From Princess Louisa Inlet, we proceeded to Calgary where I was to contact the Dominion Forest Service headquarters before commencing my summer's work in the mountains.

At the time of my marriage, I was involved on a survey of insect problems in the forests of Alberta, which I had started two summers previously. These forests extend from the summit of the Rocky Mountains (the western boundary of the province) eastward to the rolling foothills, and northward for the length of the province. On the eastern slope, devoid of roads of any kind, travel was restricted to trail by saddle and pack horse. While these trails were developed and maintained by the Dominion Forest Service, they were, for the most part, the original travel routes of the Stoney Indians, who continued to use them.

During the two previous summers, I had covered the forest regions extending from the International border northward to the Red Deer River, including all of Banff National Park. In order to do the work, I had been obliged to work in company with the ranger or warden who was in charge of the district I was visiting. He had to provide me with horses and to accompany me through his territory. It was frequently very inconvenient for him, caused me many delays and was, in fact, a very inefficient and frustrating way of doing the work. After two summers of this arrangement, it seemed to me that I could achieve a great deal more and cause much less inconvenience to others if I procured my own horses and operated quite independently. I proposed to take Vi with me.

I discussed this proposal with my chief in Vernon, and he was quite enthusiastic about it. As I was doing this work at the special request of the Dominion Forest Service in Alberta, but was in no way responsible to them, he felt certain they would be receptive to the idea.

The next morning, accompanied by Vi, I called on the Chief Forester in Calgary to explain the proposed changes in procedure for my summer's work.

"I will be working quite independently from the Forest Service this summer," I explained. "My Department thinks this arrangement will facilitate my work, and at the same time it will avoid the inconvenience to your men of having to supply me with horses and an assistant in every district I visit as was the case in the two previous summers." Having advised him of my intentions, I anxiously awaited his reaction and reply.

For a moment he sat thinking. It was obvious the suggestion

had taken him by surprise. Looking at Vi he finally asked, "What experience have you had with horses?"

"Not much, but I can ride. I can learn with experience."

Studying her further he asked, "Would you mind telling me how old you are Mrs. Richmond?"

"Twenty."

"You're not very big, about five feet, I'd say, and not much over a hundred pounds?"

"Five two and a hundred and two pounds," was her reply.

He leaned forward across his desk, resting on his elbows. "What concerns me is something of which you are not aware. If you accompany Hec, as you propose to do, you will face several extremely dangerous situations. You will have to cross the Saskatchewan River, the Southesk, the Brazeau and others which at that time will be in full flood with the summer melting of the ice fields at the higher elevations. Some will have to be crossed by swimming your horses. I feel the hazards are far too great for you to tackle in the manner you're proposing." Then, turning to me, he continued, "for you alone to tackle these situations is tempting Providence. You must have an experienced assistant in such difficult situations. I would feel responsible if, through my approval, you came to grief. I'm sorry, Mrs. Richmond, but I cannot agree with you accompanying Hec on this work."

There was little else I could say, although I began to question my wisdom in having brought the matter before him, since the Forest Service had no jurisdiction over my work nor how I travelled. Although I had the complete approval of my own Department, it seemed evident that I would have to proceed as I had done in the past two previous summers, and leave Vi until my work was completed in the autumn.

With this unexpected turn of events we decided that Vi would remain with friends in Calgary for a couple of weeks after which she would go back to Vernon and await my return at the end of the summer.

Accordingly, the following day I was taken by car into the mountains where I was met at the Ghost River by the Assistant Ranger for that district. Vi accompanied me that far, and we said good-bye for what we thought would be five months. By saddle and pack horse, I proceeded to the Red Deer Forest District station where I joined the Ranger in charge, Tom Harvey. Although we had never met, I had heard a great deal about him. He was a "loner," an individualist, a man whose reactions to any situation were entirely unpredictable. If he liked you, the world was yours; if not, you simply did not exist.

For the next ten days I travelled with him over his district, and we gradually got to know something about each other. Generally speaking, we seemed to get along just fine. One thing which stood me in good stead with him was my effort to ride a hard-to-manage and unpredictable horse. This was the only one available and had been assigned to me much against Tom Harvey's better judgment. The animal periodically went into wild tantrums, threw the rider and ran away. Tom's advice was that I should spend time with the horse and establish some kind of relationship with him before attempting to either saddle or ride him. The whole arrangement had very little appeal to me as I had limited confidence in my ability to ride a bucking horse. With the utmost patience, however, I conveyed my concern to the horse and he, in turn, probably felt a bit sorry for me! For the duration of my work in Tom Harvey's district, the horse and I worked together without the slightest disagreement.

One night, as we sat by a campfire in the dusk of the evening, Tom surprised me with the comment, "I don't think those people in Calgary have any authority to forbid you to take your wife with you if you want to."

With that, he outlined a proposal about which he had obviously been thinking for some time. He suggested he would rent me four horses which belonged to him; two saddle horses and two pack horses, plus all the equipment I would need. "With that you will be using nothing belonging to the government. It's all my private property which I will rent to you." This suggestion sounded too good to be true. As Vi was still in Calgary, I would have no trouble contacting her.

A few days later, after returning to headquarters, I stepped into Tom's storage shed to find him laying out equipment. "I'm sorting out what you will need when you bring Vi out," he explained. With that I set to and helped him assemble the various things we would require: two saddles, two pack saddles, horse blankets, tarp, canvas pack covers, small tent, bells, hobbles, cowhide pack boxes, ropes, cinches, etc. Axe, utensils, food items and so on I would purchase elsewhere.

Tom Harvey's house was on the Red Deer River. Down the valley some miles, was a ranch belonging to Dick Brown. The Browns were driving into Calgary for a few days and offered to take me in with them. The road between Harveys' place and the Browns' ranch was passable only by horse and wagon. It was arranged, therefore, that I should ride by saddle horse to the ranch, then pasture my horse while I went to Calgary. It was only an eight-hour drive. On my return with Vi, she would stay overnight at the ranch while I would return to Harveys' house on my horse, get a team and buckboard and return for my wife the following day.

I arrived in Calgary as planned and headed for the place where Vi was staying. As I walked along the street, I was suddenly startled by a voice from a passing car: "Hec!" It was Vi. Within minutes she knew why I was there, what my plans were and that she would be leaving with me as soon as possible after the Forest Service had been advised.

My first move the following day was to call upon the Supervisor in the Forest Service. He wasn't in town, so I explained the situation to his assistant. To my surprise and delight, he and the others in the office were entirely in agreement with my plan. He wondered why I had bothered to come to them in the first place, since the arrangement had already been approved by my own Department. It took the onus from their hands, as they wouldn't have to supply me with horses and an assistant in every ranger district I visited. Furthermore, it enabled me to cover much more country in the course of the summer. So, with their blessing, I gave Vi the good news and together we proceeded to purchase the things we would need for the summer. Leaving Calgary later that same day, we took the train to Olds, where we stayed the night. From there we were able to hitch a ride on a lumber truck which took us to the Browns' ranch.

With Vi staying at the ranch for the night, as arranged, I saddled my horse and rode back to Harveys' place. Next morning,

I hitched one of Tom's teams to a buckboard and returned for Vi.

This was Vi's first experience in riding in a buckboard behind two highly spirited horses. For the team, it was "homeward bound" and no encouragement was needed to get them going. At a pace just under a gallop, we rolled, bounced and bumped over ruts, roots and boulders, up hill and down, and across streams while Vi hung on for dear life! When we arrived at their home, Mr. and Mrs. Harvey greeted Vi with warm hospitality and kindness.

CHAPTER SEVEN

After spending a day organizing equipment and rounding up the horses we would be using, we were ready to commence our four-month journey. For Vi, it would be a breaking-in time, during which she would have to get used to eight to ten hours in the saddle each day. For our horses, it was a matter of getting to know us and becoming accustomed to working together.

I rode lead position on my horse "Jim." Next came our two Pinto pack ponies, "Pickles" and "Painter." Last in line came Vi on her horse "Jonas." This arrangement was intended to discourage the pack horses from getting any ideas about turning around and heading home, and was to be only a temporary arrangement until we reached unfamiliar territory, when the urge to return home would be gone. From the standpoint of the two Pintos, however, the order was final for the entire summer. Despite the fact Jonas was larger and more experienced than the others, he was never permitted to walk in front of the pack horses. If Vi and I ever tried to ride together, in front of the two Pintos, they would bite Jonas in the rear—sometimes nipping Vi in the bargain—until Jonas moved back where he belonged.

An average day's ride was 15 to 20 miles. On this first day up the Red Deer River, we travelled but ten miles and then camped for the night on a beautiful green flat beside the river. I had hoped that this would be for Vi a pleasant introduction to what lay ahead, but it turned out to be a disaster. We were plagued until sunset by literally millions of flies and mosquitoes. Never again throughout the summer were we to experience such misery from insects as on that occasion. Fortunately, the setting of the sun and the cool air brought a halt to all insect activity and I tried to convince Vi that this was, in truth, an exceptionally bad experience. Our first night was spent in the silence of the mountains, a silence broken only by the occasional howl of a coyote, which, like the lonely wail of the loon, has always seemed to me to be the music of the wilds.

During the next few days, we proceeded slowly in deference to Vi, who, at the end of each day would roll from her horse, stiff, sore and totally exhausted, and be in bed, asleep, long before darkness. As Vi became inured to riding, we gradually worked up to the point where 20 miles was a normal day's ride.

A regular routine was soon established. The actual hours of the day meant nothing. With the first streak of dawn, I would get up, light a fire and get water on to heat. While I went for the four horses, Vi would cook breakfast. When the horses were in and tied, we would sit and eat a leisurely breakfast, after which Vi would wash dishes and pack equipment while I saddled and cared for the horses. She became an expert at "hefting" the packs to be sure the weights were evenly balanced when loaded on either side of the pack horse. She also became adept at throwing the diamond hitch. The diamond hitch is, I believe, the cleverest thing anyone has ever devised to do with a rope. There are several different ways of throwing a diamond, and while they all end up much the same, I had been taught a "one-man" hitch by a forest ranger in the Kananaskis country. He, in turn, had learned it from a fellow from Montana. As the name implies, it is done single-handed, and has one special advantage. The entire complicated hitch is put on the horse completely slack, then, with one final pull from the rear of the horse, the hitch becomes tight and the pack secure. Never once in the four and a half months that we were on the trail did a pack ever come loose or require adjustment.

Fortunately for us, the summer was particularly dry and we seldom found it necessary to put up the small tent we carried. Its principal use was as a mattress on which we slept under the stars, covered with a canvas tarp and the pack covers to protect our bedding from the night dew. Frost was a common occurrence throughout the summer and it was not uncommon to find ice on the water pail in the morning.

Our objective was to leave as early as possible and travel through the day, stopping after noon at the first good camping ground where there was horse feed and fresh water. This might be two or three o'clock, or as late as six or seven p.m., depending on conditions. An early start permitted an early stop and allowed me time to do my work along the way, the official duties which were the basic reason for my being there. In addition to other forest insects, the possibility of a developing bark beetle population on the eastern slope of the Rockies was of primary concern. A survey for such potential necessitated the examination of dead or dying trees, selected at random along the trail, particularly in areas where blowdown of trees had occurred. Bark beetles of various species are always present under such conditions and if our number one enemy, *Dendroctonus*, was present, he would be found. Samples of all species found were recorded and preserved in alcohol. Such a survey had never before been undertaken in the forests of Alberta.

Since the trails were still used by the Stoney Indians, who had initially developed them, a good campground was always indicated by teepee poles stacked against trees. Most Stoneys made their own teepees, for which 15 poles were required, so the poles were carefully stored in readiness for the next traveller who might pass that way.

A great benefit to us was the fact that our horses had been raised in the mountains and knew trail work better than most men, including myself. Neither pack horse would attempt to go between two trees if there wasn't enough space for the width of his pack. Nor would they attempt to go under a leaning tree unless there was room for clearance of their loads. They knew how to ford a swift-flowing river so as not to lose their balance and be swept away by the force of the current. They would stop and paw away boulders on the river bottom before attempting another step to assure a firm, solid footing. They knew how to forage for food, even through deep snow, and never had difficulty in finding all the food they could eat regardless of conditions where a farm horse would have soon starved to death. They also knew all the tricks of bluffing their master. They knew, for instance, that if they expanded their lungs when being cinched up, the cinch would become looser and more comfortable when they exhaled. I overcame this by checking the cinches after packing and before setting out on the trail. Another trick was to lie down after the pack was cinched. Such behaviour suggests a touch of colic which, if genuine, would necessitate removal of the packs and resting the horse. When one of the animals did this, with no apparent reason for colic, I simply gave him a good spank and told him to get up and to stop being foolish. Since the bluffs never worked, they were soon abandoned.

Jonas, Vi's saddle horse, was the oldest and most experienced. His favourite trick was to hide in the bushes when it came time to be saddled. Standing motionless so as not to ring the bell that hung from his neck, he would challenge us to find him. Usually this was not too hard. If we did have difficulty, we would ignore him and saddle the other horses. When time came to depart, Jonas would invariably surrender and come forward, as he did not want to be left alone.

All four horses were terrified of bear and moose. One night we were awakened by a commotion from the horses; bells ringing, horses stomping as best they could with hobbles, and Jim, my horse, who was always tethered for the night, nickering. I went out to investigate. All seemed well enough. Speaking to the horses seemed to calm them and, seeing nothing amiss, I went back to bed. Examination of the ground the next morning revealed bear tracks, which explained the horses' concern.

The general schedule was to spend two weeks in the mountains and then to work our way toward the foothills, where additional supplies might be purchased in some small settlement, reports and letters mailed and, at times, received. We did, however, carry enough food for at least three weeks, in case there were unexpected delays. The two pack horses carried all our possessions and food, the weight evenly divided between them. Our food consisted of various staple items, some canned things, a slab of bacon and as many dehydrated products as might be available, such as potatoes, onions, fruit, eggs, milk, rice, etc. We each carried a waterproof slicker tied behind our saddles, available for immediate use in the event of a sudden downpour. In such emergencies, however, we generally found the slickers had been inadvertently stored in one of the packs.

Leaving Red Deer River, we journeyed up Scalp Creek and on the fifth day came into the headwaters of Clearwater River, which we forded. Continuing in clear summer weather and starlit nights which were regularly enlivened by a coyote chorus, we reached the Ram River on the tenth day. Two weeks northward on the trail brought us to the Saskatchewan River where the Forest Service had installed an overhead aerial ferry.

This ferry consisted of a sagging cable, stretched high above the swirling waters of the Saskatchewan River, draped over a log tower on either bank and anchored at both ends in the ground by a "dead man" or buried log. On first sight, the same thing entered both our minds: how long has that log been buried and how rotten might it be? The so-called ferry in which we were to ride across the river was a small contraption only large enough to hold two people sitting facing each other, their feet braced on a single crossbar which formed the floor. The carriage hung from the cable on two pulleys. To cross on that contraption was a decidedly scary prospect.

The Saskatchewan River was big and wide in full flood condition and when viewed from this hazardous contrivance, it was a long, long way to the other side. For the horses to cross, they would have to swim with their packs and saddles which was patently impossible.

The Ranger Station, my immediate destination, was located on the other side a mile or so up the river from the ferry crossing. We decided we would leave our horses for the present, cross on the ferry and contact the Ranger for his advice.

With the horses unpacked and peacefully grazing, we climbed the tower, seated ourselves in the carriage, braced our feet on the crossbar, and then, when all was ready, I untied the rope by which we were held, and away we went. We coasted at an almost terrifying speed down the sagging cable, but at the centre of the river, we came to a stop.

We looked down at the flood of rushing water below us and wondered just how safe was that cable and the "dead man" on which our lives now depended. From this point to the top of the tower on the opposite side, was uphill. Progress was painfully slow. I would pull on the overhead cable, move the contrivance along a couple of feet, let go and, before the carriage could start rolling backward, grab the cable again and with another pull, gain another couple of feet. The procedure was made doubly awkward as it had to be accomplished in a sitting position, and the pulleys on which the carriage was hanging were directly overhead. For this reason, if you failed to let go in time, a finger would be caught between the pulley and the cable, as I discovered on my second pull when my finger was pinched and torn open along one side. It was a very tiring ordeal.

The struggle continued, becoming increasingly difficult with each foot of progress, for as one can readily appreciate, the closer we came to our destination at the top of the tower, the steeper the grade became. There was always the horrifying thought that if I ever let go of the cable long enough for the carriage to commence its backward descent, we would find ourselves back at the centre of the river, faced with the problem of propelling the carriage and ourselves up that incline again. With brute force, pulling, holding, resting, and then pulling again, we inched our way across and eventually reached the tower on the north side of the river. While I held the contraption steady, Vi grabbed the rope and tied us securely. With a sigh of relief, we climbed down, infinitely glad to put our feet on solid ground again.

We proceeded on foot to the headquarters of the District Forest Ranger, who had been expecting me for several days. When I discussed my future programme with him, his advice was that since I could not cross the horses in the flooded condition of the river, I should return to the horses, set up camp and take a day's rest. He suggested we should then return to the Ram River, proceed to its headwaters at Onion Lake, where we should turn north until we again reached the Saskatchewan River. Here another aerial ferry had been installed at a point much safer for the horses and more practical for crossing during periods of high water. The plan would involve a seven-day trip, but inasmuch as it would take me through a lot of timber I would otherwise miss, I did not consider it wasted time.

While the horses rested where there was abundant food, we spent the following day at the Ranger Station, writing reports and letters.

A few miles down the river was a small coal mine, a store and a place where meals were served. Accompanied by the Ranger, we visited the tiny settlement. After purchasing some needed supplies, we stopped at the eating establishment for supper. The food was all put out on the table and we were expected to help ourselves to as much as we wanted. For 50 cents the supper included a delicious salad, pork chops, roast beef, potatoes, turnips, beets, parsnips, two kinds of pie, two kinds of cake, strawberries and cream, tea, coffee and milk!

We left the Ranger Station that evening, re-crossed the Saskatchewan River via the aerial ferry and spent the night with our horses and outfit.

By dawn of the next day, we were packed and away, retracing our route to the Ram River, which we reached in two days. Here we changed our course, as instructed by the Ranger, and headed for its headwaters at Onion Lake.

We had been enjoying typically good summer days, but as we proceeded up the Ram River, the weather turned excessively hot, becoming increasingly oppressive as the afternoon wore on. Ominous clouds began to roll in from the west and the high mountain peaks were hidden in their blackness. This began to concern us as the trail on which we were riding was high above the valley bottom with no prospects of food for the horses nor suitable place to camp. The weather was worsening by the minute. All of a sudden, there was a blinding flash of lightning and the whole valley shuddered from a tremendous crash of ear-splitting thunder. At the same moment, the entire sky seemed to open up and rain came down in a deluge. Lightning flashed below us and thunder echoed back from the mountain as flash after flash split the black curtain enveloping us.

Drenched to the skin, rain and wind beating our faces, we rode on, wondering when the trail would drop to the valley bottom and how far we would have to go to find good horse food and a place for ourselves for the night, possibly a dry cabin, as indicated on our map. As the afternoon progressed with no apparent slackening of the storm, we decided to continue in quest of the cabin even though it was some distance off our course. Finally, about seven p.m., we rounded the trail, and there it was, not much to be sure, but at least it was shelter. It had a stove, behind which was a pile of dry wood and kindling. That was all, but it was pure luxury on a night like that one. With the horses unpacked, unsaddled, hobbled and grazing, a fire in the stove and wet clothes changed for dry things, we were able to relax in the genial warmth and comfort of the old cabin and to enjoy a hot supper which Vi prepared.

The rain continued through the night and into the next day. With a dry place to stay, we waited for the clouds to rain themselves out, which they did by late afternoon. I did some report writing and spent the drier hours in the woods on the continuing survey for destructive forest pests. Vi did some washing and, between us, we caught a few trout for supper in a nearby stream.

With the coming of a new day, we were away, our sights set for Onion Lake, the head of the Ram River. We had hoped to reach the lake by four o'clock but not until two hours later did we find it, incorrectly marked on the map, about two miles off course. There was good horse feed, but when we looked for water, there was none. The lake was little more than a stagnant pool, shaped like a very shallow saucer, about half a mile across, and surrounded by what appeared to be a dry mud shoreline about a hundred yards wide. It was much too late to look for a better stopping-place. Camp preparations completed and the horses grazing, we decided to try to get some water from the lake. Armed with two pails, we headed across the shoreline. The deceptively dry surface was a quagmire. The closer we came to the water, the deeper our feet sank into clinging gumbo. Sinking halfway to our knees, we eventually reached the water's edge, only to find the water scarcely an inch deep.

It appeared no deeper even yards out from the shore. To attempt to wade into such a treacherous lake bottom would have been decidedly stupid. To add insult to injury, minute, wiggling creatures infested every cubic inch of water. No matter—it was that or nothing. I crouched over the water and bailed part-cupfuls, which I handed back to Vi, who poured the contents into a pail. Eventually, with two pails part full, we floundered out of the mud and made our way back to camp. Our first task was to strain the wiggle out of the water! Once it was clear of all creatures, we boiled it thoroughly. One pail was for the evening and one for morning use. For tea, it was very poor; for coffee it was worse; as plain drinking water, it was unbearable. At best, it was wet and that was as much as could be said for it. I have been where water was poor and very scarce, but never forced to use such putrid water.

The horses fared better. Rains had made the grass heavy with water and since horses do not concern themselves with washing, they had all the moisture they required.

Leaving beautiful (?) Onion Lake early the next day, we camped that night just two miles south of the Saskatchewan River. After our previous night it seemed paradise to us, as it must have to the horses. The grass was knee keep and the clear, clean-flowing creek water was home to trout. No wonder there were teepee poles leaning against the trees where Stoney Indians had camped on many occasions through the years.

By five o'clock the following day, we reached the other aerial ferry of the Forest Service. Here we met the Assistant Ranger, Clarence Earl, who had been sent to meet me and to spend the next few days travelling with us through the extremely tough and rugged country that lay ahead. His boss had been instructed to accompany me himself, but apparently expecting an aged and helpless scientist, he sent his assistant instead. We didn't require help so the assistant turned out to be no more than a travelling companion for us. As he displayed a keen interest in what I was doing, however, I took the opportunity to impart to him as much information as I could on forest pests and the objectives and methods of this type of survey. He figured he had a good laugh on his boss, who would have enjoyed being along if only he had known. Clarence Earl proved to be a particularly fine type of man and with him we experienced some of the most spectacular country and enjoyable days of the summer.

While much of the land at the head of the Saskatchewan River and the Columbia Icefields has since been opened up through highway construction, such developments were not even considered at that time.

The crossing of the Saskatchewan at the second aerial ferry consisted of two phases; first, getting the horses across, and second, ourselves. Since the grazing was superior on the other side of the river it was decided to cross that afternoon. It was a better crossing place than the previous one only in that the distance was shorter, but the water was deeper and flowing at a swifter rate. The horses had to swim.

A small corral had been built at the water's edge, where the bank of the river fell away abruptly. The idea was that with a single step into the water, a horse was over his head and thus forced to start swimming. In a matter of seconds, he would be carried downstream where his return to shore would be made impossible by a sheer cliff bank. His only way out of this was to start swimming for the opposite side which, if he was an experienced horse, was exactly what he would do. The real danger was that the horse might panic and try to swim upsteam against the current until he expended all his strength and was swept downstream, probably to his death. Naturally, it would be foolish to put an inexperienced farm horse into such a situation. A good trail horse which had travelled as a colt with its mother under similar conditions found hazards like this commonplace, knowing from the outset what was expected of him, and entering the water with his mind set on getting across.

Packs and saddles were removed for transportation on the overhead cable ferry and the four horses were crowded into the small corral. With a few slaps, spanks and some shouting, the animals plunged into the river and started on their way. Never having seen them perform in such a critical situation, Vi and I watched with bated breath and considerable apprehension. For a moment, the four aligned themselves with heads upstream, swimming against the current, but as they began to realize their situation, they one by one directed themselves toward the opposite bank. With the skill and knowledge gained through lifelong experience, they fought their way through the current until at last they found solid footing on the shallow shoreline. We sighed with relief as the horses floundered up the opposite bank and, after a few shakes, commenced feeding on the abundant bunch grass. Their bridles had been left with the reins hanging loose so they would not get the idea their day's work was done and begin to wander.

For us, there was an overhead aerial ferry, similar in some ways to the one we had previously encountered, except that this one consisted of a box-like affair large enough to accommodate three

persons along with most of our gear. The carriage could be
propelled in a standing position with the three of us pulling on the
cable. The crossing, therefore, was relatively easy. On reaching
the other side, Vi and Clarence disembarked with the equipment
while I returned to the other side for the remainder of our gear.

Our mission completed, we prepared camp for the night.

CHAPTER EIGHT

Nothing is more comforting when travelling with horses than to hear at night the occasional tinkle of bells assuring you they are contendedly grazing and all is well. Silence, especially at dawn, arouses the suspicion they may have wandered off during the night. Our herd had now increased to six, as the Assistant Ranger had brought his own saddle and pack horse.

On our first morning together, I awoke at dawn to ominous quiet, save for the sound of the flowing river, the rustling of the breeze through the branches and the songs of waking birds. I dressed in a hurry and ran to where the horses should have been, only to discover they had gone. I woke the others and the three of us went to search for the wayward animals. After tracking them for some distance, we spotted them standing on a small island in the Saskatchewan River. Why they had wandered away we could only guess. Possibly the two new horses had persuaded them to go. Perhaps they had been bothered by flies. In any case, there they were, looking straight at us but refusing to budge. Although the water between the shore and the island was not very deep, the swift current made crossing by foot hazardous. We were sure the horses would eventually return of their own volition, so for the moment we left them and returned to our camp, where Vi cooked breakfast. While we were enjoying a final cup of coffee, the welcome sound of bells greeted us as the horses came wandering down the trail, ready for another day's travel.

*　　*　　*

From the Saskatchewan River, we turned up the Kline River, camping some seven miles upstream. The next day we reached Pinto Lake, named for its patchwork variety of colours and set like a jewel among the rugged peaks of the Rockies. We took advantage of this opportunity to rest and be thankful for such natural splendour.

Arriving about five o'clock that afternoon, we were delighted to find a large new Forest Service cabin. After preparing for the night and having our supper, we decided to try our luck fishing in the lake. Also fishing on the shore was an old Indian and his squaw. He spoke some English, but she none. He explained to us his secret for catching fish in Pinto Lake. Only one spot on the lake would produce fish, he said, and that during the last hour of daylight. Using a teepee pole as a rod, the hook was baited and hung over the steep bank at the water's edge. It was essential to have someone on the bank to grab the fish once it was hoisted out, before it could flop back into the water. The old Indian fished while his squaw operated on the bank. In our operation, Vi stood on an old log protruding into the lake, while I remained on the bank ready to tackle each fish as she hoisted it out of the water. Pouncing upon the flopping, squirming fish, the two of us would slide together down the bank, coming to a halt only inches from the water's edge. The trout were all good sized, averaging about two pounds each. Within an hour, Vi had hoisted five to me, all of which I successfully seized before either fish or I slid into the water!

We were glad the next day was Sunday, when we normally rested and did nothing. We commenced the day by eating a fish each, along with a stack of hot cakes, after which we washed our clothes, tried some more fishing without success, and in general loafed and enjoyed the day. Clarence showed Vi how lots of cold water got clothes cleaner than a small amount of hot water and proved that an icy stream was the place to do washing, lacking an all-purpose detergent.

From Pinto Lake, the trail zig-zagged up the mountain side, at an incline almost too steep for the horses to navigate with the burden of packs on their backs. From "Sunset Pass," one looked out over a sea of mountain ranges which seemed to stretch to infinity, eventually fading into a misty blue. Even the horses seemed to appreciate this panoramic view, as we frequently paused to rest them in the icy mountain air. In Alpine country, August brings to life the myriad wild flowers normally associated with spring at lower elevations. One steps from a dry midsummer valley bottom into alpine meadows bursting with spring flowers.

When we stopped for lunch, the horses seemed glad of the opportunity to rest and graze on this profusion of greenery after their strenuous climb to the summit. Without too much loss of time, we continued down the other side of the mountain to the North Fork of the Saskatchewan River, where we found a good camping site and stopped for the night.

The next day we forded the North Fork intending to proceed up the Howse River, from which we could cross over and reach the Mistaya River. High water, however, necessitated a change of plan so we headed instead down the Saskatchewan.

After ten days of travelling in company with Clarence Earl, we returned to our starting point, the place where we had previously crossed the river by aerial ferry. Awaiting us was the ranger in charge, Mr. Macdonald, and his wife. We set up our tent beside theirs. About midnight, we were awakened by an unexpected cloudburst; in the midst of this deluge our tent collapsed and Vi and I were forced to claw our way out from under yards of wet canvas. While we spent the balance of the night with the Macdonalds, poor Clarence had no choice but to remain in his sleeping bag under the wet, smothering canvas of our collapsed tent.

By morning the storm had passed and the land was steaming under a brilliant blue sky. We soon had our soggy equipment drying in the sun, having hung everything out on ropes stretched from tree to tree. As the drying progressed we decided to try for a deer, but managed to bag only a few grouse.

After a final night with Clarence Earl and the Macdonalds, we bade them farewell and continued on our way down the Saskatchewan River, stopping for a day on crossing the Bighorn, a tributary to the main river.

Having set our camp for the night on the bank of the Bighorn, we walked down to a stream with our pails, to get water. I carried the large pail, while Vi brought along a small one. Coincidentally, there was an Indian couple at the stream, also getting water with a large and smaller pail, much as we were. It seemed they were discussing us in their native Stoney language, for the woman started laughing. As she couldn't speak English, her husband explained to us that his wife had taken Vi to be a boy because of her clothes and had suddenly realized she was actually a woman. Pointing to Vi, he asked, "She cross big river?" meaning, of course, the Saskatchewan. "Yes," I informed him, "she did." Pointing to her again, he said "She cross Bighorn?" Once more I assured him she had. With this, he patted Vi on her head and said, "She all same Indian squaw, all same!" As we left them, I noticed one big difference between ourselves and this couple: he carried the smaller pail of water!

Our journey continued toward Nordeg, where the ranger in charge was a man named Bill Shankland, a legend in that part of the world. A dedicated, hard-working public servant, he guarded his district as jealously as if he owned it. Anyone visiting him, particularly on official business, was accorded his almost unlimited hospitality—typical of a Scottish landowner—but woe betide you if he felt that you were taking him for granted. No one likes to be taken for granted by others, but those who spend most of their lives in the wilds seem especially sensitive where such things are concerned. Long before we reached Nordeg, we had heard rumours and stories of Bill Shankland. Not surprisingly, then, the closer we came to Nordeg and to Bill Shankland, the more apprehensive we were as to what sort of reception we would receive. Would he consider my presence an unavoidable inconvenience and my work a waste of time, or would he be receptive and co-operative? I had very mixed feelings as we approached his headquarters.

Equally unsettling was the prospect of crossing the Brazeau River, at the northern limit of Bill Shankland's district. Of all the rivers along the eastern slope of the Rockies, the Brazeau, when in flood, was among the most dangerous to ford. More than one experienced traveller, including one seasoned forest service man, had lost his life crossing that river. The problem of getting across this water had been one of the strongest arguments against Vi travelling with me on this job. I had hopes of getting some help from Bill Shankland with this phase of the operation but in light of what I had heard about the man, was reluctant to ask him for assistance.

For some time, I had been experiencing severe stomach cramps. Several men with whom I had spoken had diagnosed the trouble as "rider's cramps," the result of spending too many hours in the saddle. On the strength of their advice, I took to walking a couple of miles each day, but still couldn't shake off the cramps. No sooner had we taken leave of our Forest Service friends on the Saskatchewan River, than I had a particularly violent attack, so severe I had to sit beside the trail with my head between my knees until the pain subsided. I was beginning to suspect I had appendicitis, and worried about what would become of Vi and me if my appendix should rupture in some remote place.

Nordeg consisted of a small coal mining operation, some bunkhouses with a cookhouse, a store with a section that sold pharmaceuticals, and of course, the headquarters of the Forest Service Ranger, Bill Shankland. By the time we arrived there, I was seriously ill. The pain and cramps never let up all through that day. As the only doctor in the region operated out of a small hospital some miles away, I was advised to consult the druggist who, I was told, could concoct a cure for almost anything. The druggist suggested that, if I was adamant about avoiding an operation—which I was—he could give me something to induce nausea, which might clear up the problem. Failing that, I would have no recourse but to go to the hospital. I took the magic potion. The next day, Sunday, I was inexpressibly ill. I felt as though my insides were being torn out. If the pains kept up much longer, I decided a visit to the doctor could not come too soon.

Later that evening, however, the pains began to recede. Next morning, my stomach had settled, the cramps were gone, and I was beginning to feel like myself again. By Monday night I was back to normal and able to make plans for the continuation of my work. Either that druggist's witchcraft had cured me, or I had miraculously recovered in spite of it. (As an epilogue to this incident, a similar upset occurred some years later, resulting in an appendectomy.)

On Tuesday morning we were ready to continue. The Brazeau River was a two-day ride north of the ranger station, but Bill Shankland insisted on accompanying us and seeing us safely across to the other side. For two days we travelled with him and ate his food. Under no circumstance would he allow us to contribute anything toward a meal. "You never know, in this kind of work, when and where you'll be able to get more grub. Keep it," he warned, "you'll need it." Knowing better than to insist or argue the point, we gratefully accepted his hospitality.

On the second day, we reached the Brazeau River in peak flood, and made camp, intending to cross the river the following day.

As we were camped near a large area of blow-down timber, the result of a recent storm, I devoted the following morning to my continuing bark beetle survey. Bill went to hunt for some fresh meat since we were in particularly good wild sheep country, but returned at the end of the morning empty-handed. I, in the meantime, had found the first specimens of the *Dendroctonus* bark beetle, (*D. rufipennin*), infesting spruce. Although this species does not have the economic significance of the pine beetle, it was, nevertheless, the first record of this genus in Alberta and a rather signifcant discovery.

When we had finished lunch and completed the packing, we faced the ordeal of crossing the Brazeau River. While Vi and I stayed with the horses, Bill reconnoitered. After considerable deliberation he returned and outlined his proposed strategy.

We would enter the water at the only possible approach. Even there, the water would be not deep enough for the horses to swim, yet too deep for them to get a firm footing. In this depth of water, the buoyancy of the horse's body takes the weight off his hoofs. To add to the danger, the river bed was a mass of large round boulders. Should the horse step on one of these, he would be in danger of being thrown off balance, rolled over by the force of the river, and swept away. The horse must be allowed to pause, to feel for a firm footing before taking another step.

Halfway across the river was a gravel bar where the water was relatively shallow. On entering the water, the horse would be directed toward the upper end of this bar. The very strong flow of the river, however, would tend to force him downstream, so it would be imperative that he reach the bar before being pulled by the current past the lower end. Should he fail to make it, his chances for survival would be very remote, as there was a series of treacherous rapids only a short distance down the river. After the horses had rested on the bar, we would proceed to the upper end, re-enter the water, and continue to the other side.

Bill had pretty well covered all the details, but he still had some further advice for us in preparation for this crossing.

"Before entering the river," he told us, "you should tie one end of a rope to the saddle horn and with the other end make a loop to go over the shoulder. Should you meet with an accident, hang on to the rope, as the horse will float like a log and you will eventually drift ashore. While you will not be out of trouble by any means, you will at least be out of the water. Secondly, should you fall from the saddle, try to fall on the upper side of the horse for, provided the horse maintains his footing, he will drag you ashore."

This rather grim advice sounded a bit extreme, but actually falling from a saddle under these circumstances does happen. A strange sensation is experienced when fast moving water rushes by only a few feet from the eyes. The motion of the water may cause complete loss of equilibrium and the rider may fall. This phenomenon is by no means restricted to the novice, and the Brazeau River has been responsible for more than one such drowning. Both Vi and I had experienced this sensation and knew enough to focus our eyes on the opposite bank, rather than on the water.

Having planned everything carefully, Bill led the way. With reins hanging loose, our horses followed. Vi and I relied on the common sense and good judgment of our mountain-wise horses. With the rushing water reaching within inches of their backs, yet not quite of swimming depth, our horses carried us slowly but carefully, pausing with each step to paw aside boulders until firm footing was found. We were soaked almost to our waists, but at last we were across and our horses waded up the bank on the opposite side of the river.

Bill, having completed his mission, turned his horse and re-crossed the river alone while we continued on.

Later that same day, we came to the Southesk River which, like the Brazeau, was in flood. While very swift, it did not present the same sort of hazards. Nevertheless, we followed Bill Shankland's good advice and crossed it without incident.

A mile or so beyond the Southesk, we came to a campground where stacked teepee poles indicated it had the stamp of approval of the Stoney Indians. We stopped, unpacked, hobbled and belled our horses and prepared ourselves for the night. It was good to relax and talk over the events of the past few days. The weather was clear and bright, so we simply slept under the stars, using the tent as a mattress. It is wonderful how quickly a brightly burning fire will produce an atmosphere of comfort and security in even the most remote and lonely setting. We went to sleep that night to the sound of tinkling bells, which told us the horses were peacefully grazing and that all was well.

If only we had known what awaited us with the break of dawn!

CHAPTER NINE

My first reaction upon awaking was as always, to look in the direction of the horses. Jim, whom I staked at night, was not there, nor was there any sound of a bell. Immediately, I woke Vi. "Vi," I said, "the horses are gone!" She sat up with a start. In a matter of seconds we were up, dressed and off on a search. Jim had broken his tether and all four horses were gone. Something very unusual must have happened during the night. Normally a loose horse will head back in the direction from which he came. In this case, however, they were so far from home they would have no idea which way to go. Assuming they had probably gone on to the trail, we first looked for tracks. Sure enough, they had headed down toward the Southesk River. There on the trail was the most disturbing evidence of all. A fifth, unshod horse was with them. Clearly a wild horse had entered our camp during the night and enticed our animals away. The presence of this wild horse would make ours doubly difficult to catch, despite the fact they were hobbled. A smart horse soon learns how to run with hobbles, travelling in a loping motion with his front legs together. It is very awkward, but a determined horse can still move at a fairly fast gait, especially if he does not want to be caught. We had to catch our horses and take them from the magnetic attraction of the wild one. It was but a short time until we spotted them. At the sight of us, they stopped, looked at us in a somewhat ashamed manner while the large white wild stallion tossed his head, whirled around and plunged into the woods. Defiantly, our four kicked up their heels and loped after him.

We realized we were facing a serious problem. If the horses crossed the Southesk River, we would have no hope of retrieving them. It would be certain death to attempt a crossing of that flooded river on foot. We decided to circle around, try to get between them and the river and prevent them from reaching it by trail, and then drive the wild horse away. Our plan had almost succeeded when the stallion realized what we were attempting to do. He wheeled, galloped past us down the trail and into the river. In seconds he had crossed and was standing on the south bank, nickering for ours to follow him. Seeing how easily it could be done, our horses decided to make a dash for it. Plunging through the brush on both sides of us, they beat us to the river, and before we could prevent them, three of our horses stood beside the stallion, nickering for our fourth to join them.

The fourth, who had not yet crossed, was Jim, the boss horse. Without him, the other three horses were lost. They waited, nickering and calling to him. Jim, in his mad dash, had miscalculated. Instead of crossing where the trail led, as the others had done, Jim dashed through the woods and ran out onto a small finger of land projecting into the river. To his frustration, he discovered the river side of this little peninsula was a perpendicular cliff and that his only way across was to return and use the trail, where Vi and I stood waiting. I called and tried to entice him to surrender, but he just stood there. This was a crucial moment. If we lost Jim, we would lose all four. If I could catch him, I knew the others would return.

We waited, staring at one another, each waiting for the other to make the first move. Eventually, he decided to make a break for it. With one mighty plunge, he came directly at me. The strip of land was too narrow for him to dodge me. I made a lunge for him, threw my arms around his neck and hung on, being sure my legs were well off the ground to avoid being trampled. As expected, Jim knew the game was over and submitted while I slipped a rope around his neck. As we led him away, the other three immediately plunged into the river and followed, leaving the wild horse alone on the other side, nickering obscenities at Vi and me.

The hunt over, we returned to our camp, the four horses in tow. I tied them securely in case they got any further ideas, while Vi busied herself cooking some bacon and eggs and coffee. She summed up the situation with the comment, "Let's hope we don't have another 24 hours as tense as these last have been." We didn't—not quite!

After two experiences of horses wandering in the night, I resorted to a much-used practice of hanging a short rope from the hobbles so that if the horse tried to lope, he would step on the rope with his hind legs and trip. It proved an effective way of discouraging further attempts at running away.

*　　*　　*

We broke camp and travelled all morning over a good trail, with nothing in sight to blemish the perfect day. As we passed a few scattered blazes leading away from the trail, we stopped and checked our map before proceeding any further. The only trail on the map was the one on which we were travelling. Reassured, we continued on our way. Suddenly we emerged from the woods onto a broad expanse of meadow-like land across which the trail extended. We were soon on soft ground over which a corduroy base of logs had been laid down, making a walkway about six feet wide, a common thing where a trail runs over a boggy, wet ground. We started along this trail without concern. As we progressed, the surrounding land became wetter and wetter, and we soon found ourselves in the middle of a floating muskeg. The logs underfoot became more and more rotten; some were broken or missing entirely. Our horses proceeded with great caution, lest they should step between the rotten logs, become permanently bogged down, or even break a leg. Suddenly, Pickles, one of the pack horses, did get his leg caught. Happily, however, he was able to extract it, none the worse for wear. Then my horse Jim did exactly the same thing. He, too, recovered his balance without injury.

At this point, we stopped to consider just what we were going to do. The situation was becoming dangerous. I walked ahead to reconnoitre. Worse conditions lay ahead. It would be foolish to proceed, I decided, so we had to turn back.

Turning a horse around on such a narrow catwalk is no small feat. Fortunately for us, however, our trail-wise animals realized their predicament and were accustomed to finding their way out of difficulties. Turning them carefully, one at a time, we finally got the four horses facing in the opposite direction. Reaching solid ground again, we retraced our route to the blazed trees we had passed earlier.

69

Obviously, these blazes marked a detour around the treacherous muskeg. It was too recent an innovation, apparently, to be marked on our maps.

*   *   *

The following days were peaceful and sunny, giving me ample opportunity to concentrate on my work without the pressure of dealing with troubles on the trail. The days were getting shorter and the nights colder—not that any nights in the Rockies were particularly warm. I knew, from previous summers in the Rockies, that the autumn Indian summer was often preceded by a September snow fall. Once we were past that, we should enjoy six to eight weeks of glorious autumn weather, the best time of the year for travelling, free of mosquitoes and flies, the countryside a riot of autumn colours and flooded rivers a thing of the past.

When we reached the Luscar Mountains and called in at the headquarters of the Ranger in charge of the region, we were informed that if we continued to travel north, we would come to an empty Forest Service camp in which eight men had been staying while working on the trail. These men had been called away for fire fighting, so we were free to stop there and make use of any food with which the camp had been stocked. Although the day had been warm and sunny, the Ranger warned us that a storm was brewing. His prediction proved accurate, for as we travelled northward, the weather began to change.

By the time we reached the government camp, clouds were building up and the temperature was falling. We settled ourselves in the large cook-dining tent which contained a huge steel cook-stove, a copious supply of wood and box after box of canned food of every description. There were cartons of eggs stacked to the roof. This was without doubt the best possible stopping place should the impending storm materialize. The horses, too, found conditions ideal.

The following morning, I looked out from the tent at twelve inches of snow, and the white stuff was still falling heavily. Feeling warmly secure, I went back to bed. With a good supply of wood for the stove, we spent the day reading, writing and taking it

70

easy while the horses enjoyed the rest, pawing through the snow to graze on the abundant grass beneath.

By evening the snow had stopped, the sky cleared and the temperature dropped well below freezing. For two days we waited while the bright sun melted the snow. The trails were wet and the ground soggy but we decided to travel on anyway. I still had a lot of country to cover, and time was going by.

During the next three days, we experienced the season's most difficult terrain and the poorest travelling conditions. Wind-blown and leaning trees blocked our way as we endured periodic rains. To add to Vi's discomfort, she developed a serious rash across her chest, much like the effects of poison ivy, a problem further aggravated by the action of riding.

It had been previously arranged that an Assistant Ranger and I would meet at a certain Forest Service cabin, the Gregg Cabin. I found the place on the scheduled date, but there was no sign of the man I was to meet. We stopped the night, but as he had not turned up by morning, we decided to move on, leaving a note in case he should arrive later.

We took what appeared to be the only trail, but the further we progressed, the worse things became. To add to our troubles, it never stopped raining, and by this time, my wife was in agony. At noon, we unexpectedly found ourselves staring at the McLeod River, over which the trail apparently crossed. Our intended destination was not across the river. We were on the wrong trail. There was nothing to do but return over the torturous route. By four p.m. we were back where we'd slept the night before. To our surprise, the Assistant Ranger, Stan Scoble, had just arrived, having got the date of our meeting mixed up. We went into the cabin and made a pot of tea. After a short rest, we continued on our way in company with our host. Retracing the ground over which we had come the night before, we eventually came to a junction with a new trail which was fast and easy, and which brought us to our goal in a matter of a few hours. We learned the trail we had taken was an old Indian route which, not being used by the Forest Service, had not been cleared of windfall and debris.

Two days later we reached the District Headquarters at Coal-spur, where the District Forester and the Resident Engineer lived.

Luckily, their wives were both graduate nurses. They put Vi to bed in the Supervisor's house and brought in a doctor from a nearby coal mine operation. The three of them decided Vi should remain in bed until the rash cleared up, although the doctor doubted she would be in a condition to ride for the remainder of the season.

While Vi was recuperating, I went on with Stan Scoble to complete the programme outlined for me at the beginning of the season. Among other things, this included investigating a report of dying timber, which required a good many miles of foot work through timber and muskeg without benefit of trails. The starting point for this trek was a cabin which we reached on our second day. Here we corralled our horses safely until our return, which we hoped would be no later than a day.

By six o'clock the next morning, we were on our way. Taking a compass bearing, we set out and reached the muskeg by ten o'clock. For the next four miles, we plunged through wet ground and muskeg, sinking with each step into mud, moss and a tangled mass of low-growing brush. The day was hot, the swamp water warm and the going exceedingly difficult. We ate pounds of ripe cranberries, sour as they were, to quench our thirst. There was but one blessing—no mosquitoes, which is typical of this season in the mountains.

By three that afternoon we reached the opposite side of the muskeg, and came to the timber in question. For all this man-killing toil, the venture proved relatively trivial, but at least served to assure the authorities there hadn't been any loss of timber. The trouble, as it turned out, was a foliage blight which made the affected spruce look as if it was dying. Considering the great expanse of this blight, the concern of those who had seen it from the air was well justified.

In the late afternoon we commenced our homeward journey. Following a compass direction once again, we emerged from the forest a few feet from our starting point. It was close to ten that evening by the time we dragged ourselves into the cabin. We lit the stove, cooked two huge elk steaks, and collapsed into our beds, to be greeted in the morning with a stack of dirty dishes.

Five days after heading out from Coalspur, I returned to the

Supervisor's residence and found Vi in good spirits, recovered to the point where she was eager to be away again. What the affliction was we never discovered, but once a year for many years the old trouble returned in a very mild form.

CHAPTER TEN

I had now travelled as far north as the Athabasca River and completed the projected summer's work. My journey, however, was far from over, as I had to return the horses and equipment to Tom Harvey on the Red Deer River, a trip which took 22 days.

I had the horses re-shod, repaired some of our equipment, and after purchasing enough supplies to see us through, we left our friends at the District Headquarters and headed south. It was now the best part of the year to be in the mountains. From this time on, the autumn colours would become increasingly more beautiful, with the infinite shades of yellow, peach and red of the aspens, the fading green of the larch and the riotous colours of shrubs and low-growing brush. It was a season of deep blue sky, warm autumn sun and crisp, frosty nights, a time to listen to the melodious bugling of the bull elk across the valley bottom or on the mountainside. It was too beautiful to be anywhere but out-of-doors and we felt particularly fortunate to be in the mountains. We travelled with our horses belled to announce our approach to any fighting moose or elk that might be on the trail. Nothing terrified our horses more than a moose. To encounter a large fighting bull moose on the trail was something to avoid at all costs. On several occasions we came to places where we could see that a moose battle had occurred. In one place, there must have been a long and serious struggle for supremacy, for the ground had been ploughed and scraped by sliding hoofs; saplings and small trees were smashed and uprooted, brush and shrubs torn from the ground. It had been a very recent battle, something I would have hated to encounter with a string of horses.

Southward we travelled, by the most direct route possible. The rivers were now lower than those we had encountered earlier in the summer, and delays fewer, except when we encountered early snow.

On September 21st, we reached the Brazeau River, over which

we had crossed so perilously earlier in the year. With the water lower, it presented no difficulty. The weather did, however. By mid-afternoon, a north wind began to blow; the temperature dropped as the wind rose, and before long it started to rain. The rain then changed to snow, and soon we were riding into a blizzard. We were unprepared for such an onslaught, nor were we dressed as warmly as we should have been for such a day. We were already wet and getting colder by the minute. We rode on into the wind and blinding snow, chins tucked into our collars, the icy wind cutting our faces and daylight quickly slipping into darkness.

Ahead lay our goal for the night, the cabin where we had visited with Bill Shankland on our way north. The way seemed endless. We approached each bend in the trail with great expectations but it seemed we would never get to the cabin. Then, just before darkness, the cabin hove into view, our home for the night, our refuge from the storm.

Before we could relax, there was the job of unloading the pack horses. My cold, cramped fingers seemed to have lost all strength and I was unable to grasp the ropes tight enough to untie them. Luckily, Vi came to my rescue. While she untied ropes and packs, I carried all the equipment and saddles into the cabin and got a good fire going in the stove with the dry wood and kindling left by the previous occupant.

Once we had unpacked and unsaddled, the next most important thing was to attend to the horses. Pastured where there was good food, protected by trees from the onslaught of the wind, belled but not hobbled, they were turned loose to feed at will. They would be hobbled before our bedtime, after they had time to paw away the snow and graze. This done, we looked to our own comforts.

A blazing fire was warming the cabin as we changed to dry clothing and hung up our wet garments to dry. Vi prepared a couple of large moose steaks while the snow and wind beat against the window. We revelled in the coziness of the weather-proof cabin, enjoying our hard-won comfort.

It continued to snow through the following day until there was over two feet on the ground. I occupied myself by cutting and hauling firewood to replenish that which we had used. Vi took advantage of having access to a stove with an oven and baked a pie, using canned peaches for a filling. She also prepared a moose meat stew.

The next day the snow was deeper than ever, but we decided to travel on in spite of it since this could be the beginning of winter in that region and conditions might well get worse instead of better. Once beyond this area we knew we would be in warm autumn sunshine again. There was another cabin about 19 miles south which by its location on the map appeared to be at the extremity of the bad weather. Accordingly, we were again on the trail by 7:00 a.m. All went well until we reached the Blackstone Gap, a two-mile pass where the trail followed the mountainside through a bed of loose shale rock. This gap acted like a funnel through which the wind blew at hurricane force. The trail was totally obliterated. The only way to assure the horses of safe footing so they would not step off the trail, slip on the loose shale and roll down to the creek below, was for me to walk ahead and break trail through the freshly drifted snow. It was a windy, cold and extremely tiring two miles.

Once through the pass, and into a more sheltered spot, we stopped, lit a fire, heated some of the moose stew and made a pot of tea. While the horses rested, we enjoyed a hot lunch by the warmth of the fire.

We reached the cabin about 6:30 p.m. Once more, I was confronted with the difficulty of trying to untie the ropes with hands useless with cold. Again Vi came to my rescue.

This cabin was old and much less inviting than the one we had used the night before. There was an old rusty stove, some dry wood, but nothing more, not even a table or a chair or so much as a box to sit on. But it did have a roof, four walls and a stove, the three essentials on a stormy night. We moved in and I got the stove going while Vi prepared supper.

An early night was in order after such a strenuous day, particularly since we intended to leave shortly after daybreak the next morning in order to get completely out of the snow region. After a good wash and a hot supper, we spread the bed on the floor, extinguished the candles and settled down for a good sound sleep.

No sooner was the light out than something ran across the bed.

"What's that?" Vi asked in a startled voice.

"Probably a mouse, but he won't hurt. Go to sleep and forget about him." But within seconds, the first was followed by several more which scampered across the bed.

"Light the candles," Vi said, sitting up in bed. "We can't sleep with mice running all over us."

"We've got to get some sleep in spite of the mice," I replied. "It's been a long, hard day and probably tomorrow won't be any better. We've got to get out of this snow country and we've got to get some sleep. Anyway, the mice won't hurt you." So we slept while the mice played!

On going to bed, Vi wore a wool cardigan for added warmth. My wool socks and wool sweater were on the floor beside the bed. When we awoke in the morning we found all exposed wool had been taken by the mice during the night, to make themselves winter homes. Our sweaters were completely ruined, socks reduced to but a few shreds. Even one of the blankets had been shredded where exposed. But that was not all.

Our food was packed in a pair of cowhide paniers or "grub boxes," the top of each being closed with a cowhide flap, laced securely. These had been hung from rafters of the cabin for safe-keeping through the night.

I lowered them for Vi to open and prepare breakfast. No sooner was the lid of one open than mice poured out like clowns from a small car in a circus. They scurried across the floor and disappeared. The second panier was the same—mice, mice and more mice. There was no question what they were after—food—our food. But how did they get into the carefully stored packs, hanging from the rafters? The only possible way was for them to have dropped from the rafters onto the grub boxes. The lines of communication within the mouse community must have been operating at a very high degree of efficiency.

Their principal damage had been to those items contained in canvas bags. While not critical, this mouse invasion did create a somewhat unusual problem. The mice, in the course of their investigation into the contents of these bags, had left their "calling cards." Salvaging the food called for removal of these

small pellets. In the case of sugar, this was not too difficult. Nor
was it any special problem in the spilled rice. When it came to the
tea, however, it was something else! It called for good eyesight
and concentration because I do like a good cup of tea.

Having survived the night with the mice, we prepared to leave
just as soon as I replenished the wood we had used during the
stopover.

On the trail again, and with steadily diminishing snow underfoot, we rode through the day into warm Indian Summer weather.
Camping each night we would awaken with white hoarfrost
coating the canvas tarp spread over our bed and a thick layer of
ice on the water pail.

Eventually we reached the Saskatchewan River at the same
place where we had previously experienced such a precarious
crossing on that small and extremely dangerous two-passenger
overhead cable ferry. This time, we were crossing from the north
to the south side of the river. Water, now, was much lower than in
mid-summer and the horses could cross in comparative safety.
Swimming depth might still be encountered so we decided first to
transport all our gear across on the overhead cable ferry rather
than risk getting the packs wet. Because of the smallness of the
carriage, several trips were necessary and so after a half day's
work and the expenditure of a lot of energy, the transfer of all our
equipment was completed. It was now the horses' turn. We drove
them in and watched as they splashed their way across. Never
once did the water reach their bellies, and they walked ashore on
the other side with no more than wet legs. All our hard work on
that laborious cable ferry was for naught. Anyway, we were
across.

Thus we journeyed southward, the days gloriously warm and
sunny, the nights crisp with frost and our enjoyment enhanced by
a total absence of annoying pests. Venison we carried remained
frozen in the packs throughout the day. It was now past mid-
October. We had travelled many miles through wilderness trails,
seen sights few people have the privilege of seeing, and done
things few people have the opportunity to experience. To our
great surprise, Vi had even grown an inch in height! As we
approached familiar mountains, our horses began to sense where

they were and needed no prodding to keep them moving. At last, on October 20th, we rode into Tom Harvey's corral. We had come to the end of the trail.

Had I known this was the last time I would ever unload a pack horse, it would have been a very sad day for me. Such "last times," however, are milestones one passes before approaching the ever exciting "first times."

The first survey ever made of Alberta forests for the presence of the destructive *Dendroctonus* bark beetle, the number one enemy of the western forests, was completed. Through three summers, (1927-29) I had combed the forests of Alberta from the International border north to the Athabasca River, including the Banff National Park and the Cypress Hills of Alberta-Saskatchewan. I had found none in pine. I had found another species of *Dendroctonus*, (*D. rufipennis*) in spruce but only in insignificant numbers, in the vicinity of Nordeg.

It is interesting to note the changes that have taken place over the past 50 years. Although the beetle was recorded in restricted numbers in the Banff townsite in 1944,[1] the first record of its presence in the forests of Alberta was in 1974.

From this there has developed an extensive infestation extending over an area 75 miles long and 15 to 20 miles wide north of the International border in the Waterton Lakes National Park, including the Porcupine Hills. It occurs also in the cities of Lethbridge and Medicine Hat. An extensive infestation resulting in the killing of pine occurs in the Cypress Hills of Saskatchewan. It is believed the insect was carried across the prairies by the prevailing winds from the west, a distance of 170 miles.

The spruce beetle, (*D. rufipennis*), present as recorded in the initial survey, is now distributed in minimal numbers throughout most of the spruce forests of the province. In only one area, however, has it become a problem of economic importance, that being in the Crows Nest region. This, it is thought, had its origin in an extensive blow-down in 1974. Moderately low populations have persisted since that time. This beetle, unlike its counterpart

[1] Hopping, G. R. and Mathers, W. G., 1945. "Observations on outbreaks of the mountain pine beetle in lodgepole pine in western Canada." *Forest Chronicle*, Vol. 21, pp. 1-11.

in pine, normally exists in small numbers in weakened or wind-thrown trees, slash or fresh stumps. Only occasionally does it arouse economic concern.

Alice and Thomas Richmond, parents of the author, photos taken in the late 1800's.

Thomas Richmond, author's father, age 18, riding for ZH Ranch, Muskateen
Cattle Company, Oklahoma, 1885.

Riders of the ZH Ranch on round-up, Oklahoma, 1885.

Author's sisters and brother, from left to right: Donella (Ella); Tommy; author; Gwen and extreme right, Alice, 1904.

The famous "Lorna" trestle, Kettle Valley Railway, east of Penticton, B.C., 1928.

PHOTO: R. N. ATKINSON MUSEUM, PENTICTON, B.C.

Vi, on honeymoon at Banff, Alberta, 1929.

The author, packing the two horses, "Pickles" and "Painter," 1929.

Pack outfit in the Rockies of Alberta, 1929.

Crossing the Sunset Pass, altitude 9,500 feet, Canadian Rockies, accompanied by Clarence Earl, Assistant Ranger, 1929.

September snowfall, Rockies. Vi on her horse "Jonas," 1929.

Tin heater installed for the comfort (?) of Mr. E. C. Manning, Chief Forester from Victoria. Vi, seated in the tent; the old Maxwell parked alongside, 1930.

Vi and author in a temporary camp while doing field work, 1930.

River boat and party during the survey of Tweedsmuir Park, prior to building the Kenney Dam and the flooding of the drainage system through the park. Personnel shown: at stern, Frank Hanson, guide and outfitter; left side, Dr. Dave Turner, Dr. Pete Larkin and the cook; right side, Rus Potter and Si Oldham.

"Pansy"

The flooding of Tweedsmuir Park, B.C., 1950.

Derelict reminders of long-lost dreams of early settlers, Queen Charlotte Islands, B.C. A crumbling home. Jim Kinghorn in doorway.

An oxen yoke, a reminder left by an early settler, Queen Charlotte Islands.

An old kerosene, single cylinder tractor abandoned in the woods by a departed settler, likely dating from before the first world war. Queen Charlotte Islands, B.C.

Abandoned farm on Queen Charlotte Islands. Collapsed roof, rotting floors, fallen fences and acres of what were once cleared land remain as testimony of the ambitions and dreams of early pioneers.

The survey vessel, *J. M. Swaine*, operated on the B.C. Coast by the Forest Insect Research Laboratory, Victoria.

Protecting logs from attack by wood-boring insects, without insecticides, using a water mist produced by high pressure pumping system (B.C. Forest Products operation).

Old growth, Douglas fir, B.C. Coast.

Typical Ponderosa pine stand, B.C. interior dry belt.

Richmond residence overlooking the sea on Vancouver Island, B.C.

CHAPTER ELEVEN

I learned the hard way about the risks involved in buying a used car, especially an obsolete model, and particularly so if it is also already worn out. I did not know its condition, of course, when I bought it.

The year after my work in the Rockies had been completed, I planned to return and continue my survey work of the three previous years, but due to the impending transfer of the forest resources of Alberta from the dominion government to the province, work plans were changed. It was decided I should continue this survey through the railway belt of British Columbia. The railway belt was a strip of land extending the length of the railway to a width of 20 miles each side of the line. Forests within the railway belt were under the jurisdiction of the dominion government.

For this work, I needed a car. The government was unwilling to provide me with one, but offered me mileage if I would provide my own. This seemed reasonable if only I had a car or the cash with which to buy one. Most of my earnings thus far had gone to pay off college debts.

After some searching, I found a car that seemed to fit the bill. It had scarcely a scratch on it, no dents and the engine ran smoothly. The price was $125.00. It was a 1923 Maxwell touring car. I took it to a mechanic highly recommended by a friend of mine, who pronounced it a very good buy and said with a few repairs, he could make it "as good as new." On his advice, I purchased it, then instructed him to bring the car up to scratch. When I returned, I found the repairs included a new battery, new universal joint, brakes relined, (two-wheel brakes of course), repairs to wiring, new king pins and bushings in the front, a new bearing in the transmission, a new speedometer cable, headlights welded, carbon cleaned, valves ground and a new tire and tube. The total bill was $60.00. The seller of the car, on learning of this big repair bill, dropped his price from $125.00 to $110.00. So, after paying the repair bill, I drove away with a car "as good as new" for just $170.00.

Vi and I were very proud of this acquisition. It was our first car, the first big purchase we had ever made. It started easily and ran smoothly, but it did use a lot of oil. In fact, we had to add oil as frequently as we did gas. This was of some concern to us, so I returned to our mechanical wizard and told him of the trouble.

"No problem," he assured me after a close inspection. "All you need are new piston rings. Once those are installed, it will be as good as new."

I'd heard this promise before, but resigned, took his word for it and left the car to have new piston rings installed.

Once the operation was underway and the insides examined, he phoned me again. In a rather apologetic voice, he announced, "I have run up against another problem. The new rings won't be enough, you see; the cylinders themselves are badly worn."

"And that means?"

"I guess it'll be a re-bore job," he confessed.

"Oh Lord!" I exclaimed. "And how much is that going to cost?"

"Well," he very meekly admitted, "You see, a re-bore job won't be quite enough either. It's going to take new pistons as well. And this is an obsolete car; they will have to be made to order in Vancouver." Then, changing to a very cheerful voice, he assured me, "But this will give you an almost new engine."

Eventually, the job was completed. We scraped together the necessary money to pay the bill and as I took the car from the garage he gave me one final bit of advice: "Take it easy for the first few miles until the new pistons get worked in. The motor may heat a bit, but don't worry. That's to be expected."

With that warning, I drove away, very carefully, but within five miles, the motor was boiling like a tea kettle. In fact, it boiled all summer long—winter too—every time I ran the motor! I doubted if those pistons would ever get worn in. One thing in its favour, oil consumption dropped to nil. I wondered if the pistons were getting any oil at all.

The season's work was going very well. I located a number of very serious bark beetle outbreaks in pine, fir and spruce, with a lot of dying or dead timber. An extremely large outbreak of the tussock moth was sweeping the Adams River country, although most of the timber eventually recovered. Meanwhile, the old Maxwell was fulfilling the mechanic's prophecy. It ran without fail, so long as a boiling motor was not considered a hindrance. I was working up the North Thompson River at the time and the narrow, winding gravel road extended as far as Vavenby. I was scheduled to meet the Forest Ranger at that small community. As we approached the settlement, something very queer happened to the motor, accompanied by a loud internal clattering noise. The motor gave one final, agonizing gasp and died. I coasted down the grade into the settlement and came to a stop.

The operator of the one and only gas pump was, by his own admission, no mechanic. I had no alternative but to do something myself.

It was quite clear the trouble was internal, so to start with, I removed the oil pan and looked inside. As soon as I did this, a lot of pieces of broken metal fell out! Now I saw the trouble; one of the pistons had broken like a china cup. To extract the broken piston meant removal of the piston rod and a few other things. To do that, however, I was going to have to take off the cylinder head. This was becoming far too big a job to finish before I had completed my mission with the Forest Service, so I travelled with the ranger in his car until we finished, at which time I returned to work on the Maxwell.

By good fortune, the CNR had a work train parked on a nearby siding and Vi had made friends with some of the crew. They had, in fact, invited her to have dinner on the train. In the course of this acquaintance, they learned of our plight. Since there was no motor mechanic in the district and tools suitable for working on a gas engine were almost non-existent, the entire work crew arrived after supper armed with every conceivable type of wrench and spanner. All of the tools were designed for work on heavy steam equipment. None of the men knew anything about an automobile engine, but all were eager to help. Since I, too, lacked expertise, it was a very agreeable work party!

We continued from where I had left off, extracting the rest of the broken piston, disconnecting the piston rod from the crank shaft, then turning to the removal of the cylinder head. It was a difficult job, but with the combined skills of the members of the train crew, the head and gasket were eventually removed and the last of the broken parts extracted. We replaced the gasket and cylinder head, secured the bolts and replaced the oil pan and the oil. We then tried the motor which was now shy one piston. It ran, a bit ragged to be sure, but at least it would get us back to Kamloops where we could have it properly repaired. With a farewell and thanks to the train crew, Vi and I headed for Kamloops, despite the late hour.

What none of us realized was that with one piston missing, we completely upset the vacuum feed of the gas line. This could have been avoided, had we known, by sealing the opening of the missing piston with a sheet of heavy paper before replacing the cylinder head. Because the fuel no longer flowed to the motor in its normal way, the car would only run for five minutes before the motor would stall. This would be followed by a trickling sound of gas running into something or other. When the trickling sound stopped, the car would run for another few minutes, then stall again.

Another unexpected problem was that the headlights were not very good. They were so dim, in fact, that had we been travelling any faster than a snail's pace, we would have inevitably run into the ditch. We had probably loosened a wire while removing the broken piston.

We continued, one stuttering mile after another, until, rounding a bend in the road, we came on a herd of horses on the roadway. All but one of them ran off to the side. That one, annoyed or hypnotized by the headlights, dim as they were, spun around and then, facing the car, charged directly into it. For a moment, it appeared he might come over the hood of the motor and land on the fabric top of our touring model. I swerved the car, hoping to miss him, but he hit the right headlight, smashed it, and at the same time crumpled the right front fender back to the axle. Then, with one mighty leap, he cleared the engine hood, sailed through the air and landed on his four legs on the roadway. At that moment, he let fly with his hind legs, giving the rear of the car a mighty kick which smashed in the back of it, driving the fender into the tire. We got out to survey the damage. Since the horse had disappeared into the darkness, we assumed he, at least, was still functioning normally! The car would still run, although it looked even more decrepit, with the front smashed in, one dim headlight and the rear of the car crumpled. We continued our limping journey, arriving in Kamloops just as the garage was opening its doors for a new day's business. We left our vehicle for repairs, sought out a hotel where we could get a good breakfast, a room, a bath and some much needed sleep. Little we knew that in the very near future we would return to the same hotel and the same garage for more emergency repairs.

Eventually, the car was fixed, four pistons again functioning. The Maxwell was operating normally aside from the never-ending boiling of the motor. There was, however, another defect I had considered of minor importance when I made the purchase. The starter would not work. The ring gear on the flywheel was shot and, due to its obsolescence, could not be replaced. The car had to be cranked by hand and this was an operation fraught with danger. The crank was bent and if it made a complete circle, the handle would hit the radiator as it came around. Cars of that vintage had no protective grill in front of the radiator, as do vehicles today. It had a hand throttle on the steering column as well as a foot accelerator. Before cranking, it was necessary to open the hand throttle, closing it once the motor was running.

On this particular occasion, we were many miles back of the headwaters of the Tranquille River. I cranked the motor in the usual way. The motor started all right but the crank failed to disengage itself from the crankshaft. Instead, it spun around at the speed of the motor which now was running, throttle wide open. Spinning at such a rate the vibration caused the whole car to shiver and shake until it became only an indistinct outline!

Both of us stood in front of the car watching, not daring to walk past that spinning crank lest it should leave the shaft, fly through the air and strike one of us. Eventually, it came free and hurtled through the air a hundred feet or more. It was now safe to return to the car, shut off the motor and survey the damage done by the crank.

The radiator was totally squashed in where the spinning crank had gouged it and water was squirting out in all directions from the many breaks. The immediate question facing us was how to drive from there to Kamloops over many miles of winding mountain roads in a car whose radiator would hold water for no more than four or five minutes.

Fortunately, we had with us two four-gallon gas cans, a canvas water pail, and several large pots and pans. By good luck there was a stream nearby. We filled the two four-gallon cans to carry in reserve and with the pots we filled the leaking radiator. Again we started the motor and took off as fast as possible hoping to come upon another convenient source of water from which we could replenish the water supply before our entire reserve was used up. At first, we added oatmeal and flour to help plug the leaks. That exhausted, we resorted to mud which we mixed with the water. This helped, but as the day progressed we became increasingly plastered with mud and soaked with water. The closer we came to Kamloops, the greater the difficulty in finding water. Twice we had to climb to the bottom of a steep canyon and eventually we had to crawl through fences and bail water from irrigation ditches. Our last source provided sufficient water to enable us to limp into Kamloops eight hours later, our clothes and shoes soaked with water and mud, our appearance in keeping with the derelict condition of the Maxwell. We went back to the same garage and hotel we had visited previously.

With the leaks finally closed, the radiator again held water, but circulation within the cooling system was almost blocked. This was no great worry since the motor had boiled ever since it was over-hauled at the time of purchase. Now it just boiled a little sooner!

My survey work within the railway belt was now completed. For the remainder of the summer I undertook a survey of the tremendous bark beetle outbreak in the Nicola-Aspen Grove country, establishing my headquarters in the Aspen Grove region between Merritt and Princeton. There was but one main road, rough, winding and of course, gravel. The few side roads had been constructed many years previously and either long since abandoned or seldom travelled by anyone. Due to years of use by range cattle and no maintenance, the roads carved out along hillsides had been so badly filled in, there was little left to suggest they had ever existed.

We were travelling along one of these hillside roads, the car listing at a precarious angle, when suddenly the rear end of the car dropped with an alarming jolt. Vi's first note of alarm was, "What's happened?" and almost the same breath, "What's that rolling down the hillside?"

I looked. Sure enough, there went our outside rear wheel, rolling like a hoop down the hill, gaining speed as it went, bouncing higher and higher as it hit rocks and logs. It finally came to rest at the bottom of the canyon. With a feeling of despair, we considered our predicament: a broken rear axle, a stalled car and our wheel at the bottom of the canyon. It was a long way down, and packing a wheel and tire up a precipitously steep incline is not the easiest thing to do. Nevertheless, it had to be done and after much exertion and sweat, we did it. Leaving the car on the so-called road where it had come to a stop was no hindrance since the road had not been travelled in years. After a rest, we set off for our camp, several miles out to the main road and miles beyond.

By late afternoon, we wandered into the camp, having stopped at a nearby ranch to phone to Merritt for a new axle. Fortunately, we were able to procure one at a cost of $8.25, and in due time, with the aid of a local friend and handyman, we returned to our crippled car, made the repair and once again were mobile.

Through the remainder of the season, the old Maxwell boiled along with but two further misfortunes. The first was a burned out valve; to replace it and have the others ground cost $4.40. The second was another broken piston, just as we returned to Vernon at the end of the season. That finished our friendship with the Maxwell. I traded it in the following spring on a brand new 1931 Model A Ford. (This was the last year the Model A Ford was produced.)

Financially, the year's experience had not been as bad as one might expect. The mileage reimbursement covered the cost of repairs and operation, but contributed nothing toward the investment. As a trade-in on the new Ford, however, it brought $150.00. The price of the new Ford was $827.00. Thus, with the trade-in, the new car cost a net amount of $677.00.

CHAPTER TWELVE

To me, the ponderosa pine forests of central British Columbia are the most beautiful and interesting of any in the province. The trees are widely spaced, the ground carpeted with grass and generally free of undergrowth. The surrounding rolling hills, spotted with numerous small lakes, are green and flower-bearing in the spring and brown in mid-summer, the resinous smell of pine scents the heat of the day, and at night, the coyotes yap-yap back and forth or sing their traditional chorus. We loved nowhere more than we loved that country.

Through the centuries, the ponderosa pine had grown and thrived, some attaining a diameter of more than five feet. In their maturity they were ideal as hosts for the destructive *Dendroctonus* bark beetle, which swept through this country during the 1920's and 1930's. Despite massive control programmes attempted by the provincial government and the expenditure of many hundreds of thousands of dollars, the insect continued on its destructive mission until practically all the old growth pines were killed, making way for young, vigorous pine in nature's never-ending rotation within the forest.

For the next five years, my life and work was devoted entirely to research on the habits and destructive activities of this bark beetle.

I spent the latter part of the summer and fall of that first year (the same year in which I was in the railway belt) in an effort to ascertain the distribution and intensity of the outbreak. Since my work during the latter part of the season would be of a very transient nature, I established no permanent cold weather camp. Instead of a closed tent, capable of being heated, I used a reflector tent, 7 x 9 feet of floor space and a two-foot rear wall. A canvas fly attached to the front could be lowered to enclose the tent or to use as a roof for a porch. In this type of tent, the warmth of an open fire is reflected into the tent so it can be comfortably warm without the inconvenience of a stove. This being extremely dry country, we found the reflector tent ideal. As the work progressed, we finally established ourselves on the ranch land of Jim McAvoy, who lived with his mother, Mrs. Turner, his sister, Mrs. Walker, and her five children.

Mrs. Turner was truly a wonderful woman and a delight to know. Her father was an early immigrant from England who came to this country in the mid 1800's, settled in the vicinity of what is now called "Minnie Lake," and married an Indian woman. A baby was born, named "Minnie" (who later became Mrs. Turner). In her honour, the lake was officially named "Minnie Lake." Mrs. Turner told us she was the first white baby born in the Douglas Lake country.

Jim, her son, was a man of unusual talents. In addition to running the ranch and a trapline during the winter, he did exceptionally fine rope and leather braiding, and, of all things, very fine needlework. He developed these hobbies while confined in a trapper's cabin during prolonged winter storms. One day, he showed Vi and me some of his needlework which I am sure few women could match. He was reluctant, however, to display this ability and usually kept his work carefully locked away.

Living, as we were, in close proximity to this family, we were soon accepted into the community life of the residents, a life quite different from anything Vi had known before. One day, on returning to our camp after working back in the woods for the day, I found Vi on top of a load of hay and, for the first time in her life, driving a team of horses as she helped them put up their hay crop. At that time, Jim McAvoy and his family were all living in tents while their new log house was under construction. It was due to be completed soon after our arrival and plans were already being made for the big housewarming. As the day for this gala occasion drew near, Vi became an active participant in the elaborate preparations which were underway. A piano was moved into the house and benches installed around the walls. The house was to be otherwise unfurnished until after the party to which everyone in the countryside was invited. Vi and I were looking forward to the party as a new experience. Some of the guests arrived in late afternoon in time for the evening meal and the numbers continued to increase as the evening progressed until we began to wonder just how many people the house could hold.

Music for square dancing was provided by the piano, a fiddle and guitar. Of the musicians, the fiddler, Jack Thynne, was in a class by himself. He played with the fiddle tucked into his hip. The dexterity of his fingers left no doubt he had once been a well-trained violinist.

We learned he had been born in Cornwall, England in 1865, the second son of the Reverend Arthur Thynne. His grandfather, Lord John Thynne, Sub-Dean of Westminster Abbey, officiated at the coronation of Queen Victoria. Jack immigrated to Canada at an early age, later married the daughter of a Hudson's Bay factor and came into the Nicola country in 1887. Travelling by saddle and pack horse, he brought his bride to Toulameen where he settled. Selling out in 1921, he moved to One Mile Creek, where he lived when we met him. His features and beard resembled that of King George V, to whom he was related. Jack died in 1943.

With Jack Thynne playing the fiddle, his daughter the piano, and a guitarist completing the trio, the dancing became more uninhibited as the evening wore on. Jack's daughter, the pianist, had been educated in England and presented at court, but had chosen to return to the Nicola country where she was married and lived with her family. Food was served all evening and the dancing stopped only long enough for the musicians to have a cup of coffee and a sandwich and to get their breath. The housewarming party continued far into the next day, although Vi and I slipped away with the coming of dawn. It was a never-to-be-forgotten housewarming, a night of unrestrained hilarity and fun.

As autumn advanced, the days became cooler and the hours of daylight shorter. Eventually it became necessary to add anti-freeze to protect the car during the freezing nights. We became acclimatized to the cooler temperatures and in spite of having only a reflector tent, felt no great discomfort. My work was now practically completed for the year and I was ready to leave when word came that I was to have an official visitor from Victoria. Mr. E. C. Manning,* Chief Forester of the B.C. Forest Service, was

* Mr. Manning was later killed in the crash of a TCA plane near Armstrong, Ontario. Manning Park was named in his memory.

coming to spend two or three days with me in order to see the extent and severity of the beetle outbreak.

He arrived with a sleeping bag, expecting to find a well-established, winterized camp. Instead, I had no more to offer him than a lean-to tent with no stove or door, heated only by an open campfire. Having come directly from an office in Victoria, he found this unendurable. I couldn't blame him. The forest ranger, Bob Little, who had driven the Chief Forester out from Merritt, found a canvas tarp that could be draped across the front and one side of the porch-like awning of the tent and an old air-tight tin heater which we installed for warmth. This meant we could share meals with our distinguished guest, although, for his sleeping comfort, arrangements were made at a nearby ranch.

For the next two days, I drove the Chief Forester through the region and showed him the devastation caused by the bark beetle. It soon became apparent to him that any further attempt to control the insect or to salvage the enormous volume of dead and dying timber was hopeless. The lack of knowledge not only of the biology of the insect but of how to manage a forest in the face of such a destructive infestation was equally evident.

For three nights we enjoyed barbecued grouse, cooked over the open blaze. Eating by the fire and sitting through the evening in the warmth of the tin heater made this an occasion which, he said, he thoroughly enjoyed.

My field work for the year ended with his departure and I returned to my winter programme in the Vernon laboratory.

CHAPTER THIRTEEN

With the coming of spring, we returned to Aspen Grove to commence a long-term research programme on bark beetles set up for me by Dr. J. M. Swaine, of Ottawa, one of the foremost authorities on the subject.

Studies covered many facets of the insect and its problems. The beetle's habits and biology; its reproductive rate from year to year, and the effect of overcrowding of populations on egg production; the importance and biology of its parasites and predators and the importance of predacious birds such as woodpeckers and others; a study of the tree and its rate of death from the moment of its first attack—all these were matters for research. The subject of applied control did not enter the study, since it had been shown to be generally impractical. Knowledge gained from these studies would contribute to the future development of any plan of management designed to cope with bark beetle attack. The programme for the year ended each fall with a general survey of the year's spread of the infestation.

There is a peculiar thing about the pine bark beetle—its mode of attacking a tree. Its habit is to attack en masse during the latter part of the summer afternoon. This behaviour benefits the insect on two counts.

In the first place if beetles attack only in small numbers (as happens when populations are small) they are easily drowned out in a healthy tree by the copious flow of sap from the tree. On the other hand the invasion of hundreds of thousands of beetles simultaneously soon wears down the tree's resistance, reducing the flow of pitch and increasing the success of the onslaught.

Secondly, a late afternoon invasion allows the insect about twelve hours to penetrate the bark, during which time it is relatively safe from external enemies such as birds and other predators.

A most dramatic demonstration occurred one afternoon when

a large mature ponderosa pine standing beside our camp was hit at about four o'clock. Vi was the first to notice it and called me from other work on which I was occupied. It was a spectacular sight. The trunk of the tree was swarming with literally thousands upon thousands of adult beetles, all actively engaged in chewing their way through the bark to the inner wood. The entire swarm had arrived within an hour. By dawn the following morning, the entire population had penetrated the bark and were no longer visible.

On each of the beetle's wing covers (the elytra), there is a file or rasp-like area. When the wing covers are folded over the body, the two roughened surfaces overlap. Motion of the abdomen causes these two surfaces to rub together, producing a distinct, audible chirping sound. This occurs when the beetle is in the process of making a new home and serves to announce that it is looking for a mate. When hordes of these insects attack simultaneously, the thousands of stridulating organs can be heard distinctly for a distance of ten or twenty feet from the tree.

The establishment of a camp and study centre involved a lot of work and planning. Since I was to be there for four summers, I selected an open, park-like spot, among large, mature pines and with a good fresh water supply. The camp consisted of a kitchen tent connected to an office-livingroom area and two separate sleeping tents. All had wooden walls and floors. A homemade shower bath was constructed near the water supply. For this, a large metal tank was mounted on an eight-foot tower. From the bottom of the tank, a hose line carried the water to the shower room constructed at the base of the tower. We filled the tank in the morning, allowing time for the sun to heat the water through the day. It made an exceedingly good shower bath.

Field work commenced in early spring, as soon as snow conditions allowed travel through the woods. To assist in the establishment of the camp and the construction of field experimental facilities I was authorized to hire four workmen for a period of three weeks. One of the four was supposed to be a cook. Vi offered to do the cooking (with no pay, of course), in order to give me four able-bodied men to do the heavy work in the woods. This proved very fortunate, for without that fourth man, we would never have completed the necessary work in the allotted three weeks.

With the termination of the extra labour, I was alone except for Vi, who worked with me as an unpaid assistant. She had become extremely knowledgeable on the subject and was fully aware of the nature and objective of the work. This was a good deal for the government and ourselves, since they were getting two for the price of one, and we were able to share a life and work we both enjoyed.

During this period, the depression years of the 1930's, no extra help or assistants were ever employed except under extreme circumstances and research projects were pursued single-handedly as far as possible. During the second summer, work had mounted beyond one person's capability and a full-time assistant was assigned to the work for the summer. He was Ken Graham, who later became a professor of Forest Entomology at the University of British Columbia.

Although the camp was adjacent to the main Aspen Grove road, there was little traffic, never more than three or four cars a day. Tourist travel was almost non-existent. There were, nevertheless, a great many wandering, unemployed men along the road, mostly friendly, interesting people with no place to go, still hoping that someone someplace might offer them a job, although they knew the chances were very slim. Our nearest neighbours were the Thomsons at SX Ranch, about three miles north; Jim McAvoy, four miles south; the McKenzies, seven miles south; and the Menzies, about three miles away in the Kane Valley. Since my work involved many areas throughout the Aspen Grove and Douglas Lake country, we gradually became acquainted with most of the ranches and residents of the region.

Despite axle-deep mud in the spring, ruts, dust and rough roads in the summer, it was a wonderful time to enjoy the province. Since most country roads were the same, nothing better was expected. We were free to enjoy every lake, every beach, every stream, without interference from others and without "NO TRESPASSING" signs. Vi and I enjoyed four very happy years. Winters were spent in the laboratory in Vernon working on the data obtained through the summer and spring work resumed as soon as roads permitted. Invariably, snow had to be scraped from the tent platforms before camp could be erected in the spring.

There is a plague in that country, however, that can almost drive a person to distraction until he learns how to cope with it— MOSQUITOES!! The use of repellents rubbed over the face and exposed skin are more offensive to me than the mosquitoes. Of all such materials available at that time, Citronella Oil was the cleanest and best. Although Vi suffered from the pests longer than I did, we both, in time, developed an immunity. Working outside, we would become covered with them, but once we washed with a strong soap and dried with a rough towel, all evidence of their bites would be eliminated. One standard defence against these insects inside a tent is the burning of pyrethrum. A tablespoon of this powder smouldering on a saucer releases a faint aromatic smell and a wift of smoke comparable to a cigarette. Almost at once, any mosquito in the room or tent will drop to the floor or otherwise disappear. We burned this throughout the mosquito season with no more ill-effects than a slight soreness in our throats toward the end of the summer.

Soon after we had set up our camp that first spring, a rider by the name of Jasper came into the camp, a sack hanging from his saddle. He had come from a Douglas Lake Cattle Company's "cow camp" where they were herding cattle onto the summer range.

"A present for you," he announced, handing the sack to Vi. She opened it and out jumped a very pretty kitten. "It was born in the cow camp and since we are moving and can't take it, thought you might like it," he commented.

He had certainly gone to a lot of trouble on behalf of the kitten, so Vi accepted it with pleasure, although I am not an ardent cat fancier. We named her "Tsugae," the generic name for hemlock. Although there was no hemlock in the country, the name sounded suitable.

The first few months of Tsugae's life were spent totally isolated from any other cat, and the only dog she ever saw was our Airedale, Peter, who became her very special friend. In this rather unspoiled and primitive environment, she developed a personality quite different from any city-raised cat.

One thing which distinguished her from any other cat I have known was her love of tobacco. She had, in fact, a real tobacco habit. Just how this originated we never really knew, although we learned that, as a kitten, she had been fed milk in a "snoose" (Copenhagen Snuff) tin at the cow camp where she was born! We figured this may have created in her a lust for tobacco. Both Vi and I smoked at that time (a habit we have long since kicked) and each evening Tsugae would hop on my knee if I had an open tobacco tin, take a mouthful of tobacco, hop down again and sit, leisurely eating her day's quota. Whenever we were absent for a couple of days, she would be ravenous for it on our return.

Another of her peculiarities was her apparent love of music. She could be attracted anytime, day or night, simply by turning on the car radio and leaving the door open. Tsugae would be found stretched out on the car floor listening to the radio. When in town during the winter, her favourite spot to sit was on top of the radio.

During her third summer in the woods with us, Tsugae had a batch of kittens. Prior to their birth, she disappeared and for several days we saw nothing of her. Although food was placed out and we called her name, she remained hidden from view, probably in some old rotten log or stump. By coincidence, Vi's dad drove into camp and on learning of the cat's disappearance and knowing her attraction to music, he turned on his car radio, leaving the door open. In half an hour, we checked and, as expected, there she was, stretched out on the floor, her head by the radio, listening. With a little encouragement, she escorted us to a hollow stump where her family was hidden. We carried four kittens back to the camp, and placed them in what we thought was a warm, soft bed behind the stove, but she carefully carried them, one by one, to a place under the tent platform where she kept them.

Although she was a great hunter and kept her kittens well supplied with mice and rodents, she cultivated a special friendship with a rabbit. Each day she and the rabbit could be found together. The rabbit, in turn, became quite friendly with us, seeming to assume that if we were safe company for his friend, the cat, we should be safe for him. Tsugae was quite a cat, indeed!

If you live close enough to a wild animal, it eventually becomes a friend. During our stay in Aspen Grove, a little squirrel lived in a big ponderosa tree beside our kitchen tent. He knew we would do him no harm and so became quite friendly. For his own amusement—and ours too—he teased our Airedale dog, Peter, who had the habit of sleeping during the day at the base of the tree. The squirrel's game was to come down the tree, inch by inch, and with each step, stop and chatter at the dog. Peter, for his part, lay feigning sleep but was actually wide awake and watching. Ultimately the squirrel reached a point the dog considered grabbing distance, and he would leap after the teasing squirrel which was, by this time, well on his way up the tree and beyond danger. This afforded the squirrel daily fun and the dog a constant, frustrating challenge.

The squirrel had another game of hide-and-seek to which he often enticed Peter. Sitting at one end of an old log, he would chatter to attract the dog's attention. Peter invariably took off in chase. The squirrel would scoot along the upper surface of the log, while Peter bounded along on the ground beside it. Just as the dog reached the point where success seemed assured, the squirrel would drop down on the side of the log away from Peter, reverse direction and return from whence he had come, leaving a completely bewildered dog. The game was repeated on several occasions until Peter finally got wise to what the squirrel was doing. The little animal had underestimated Peter's intelligence. One day, the chase commenced as usual and the squirrel dropped from the upper surface of the log to the side, just at the crucial moment in the race. This time, however, the dog was prepared. He leaped over the log and there was the squirrel, just inches away. The race continued, but now it was a race for life.

Instinct told the squirrel the safest refuge in such an emergency was up a tree. Listening to the voice of instinct, he left the log, dashing across the grassy flat in search of a good tree. But there wasn't one nearby. The dog was closing in; the squirrel had to do something in a hurry. All he could see, other than grass, was me, standing and watching. In desperation, he raced to me and, with a single leap, landed on my chest. Then the merry-go-round was really on. Peter leapt, barked and raced around me. One moment the squirrel was on my shoulder, then across my chest to the other shoulder, on my back, over my head, onto my chest, and with each step his little needle-like claws pierced through my shirt and into my skin. I was kept busy trying to shake off the squirrel, trying to grab the dog, yelling at both of them, and calling to Vi for help! Vi, meanwhile, was doubled up with laughter at my predicament with the squirrel and the dog. Eventually, however, she came to my rescue and dragged Peter away. The little squirrel dropped to the ground and scooted across the grass to the nearest tree. Whether he had raced to me hoping I would save him, or whether he thought I was a stump standing there I was never sure. In any case, he did the right thing.

CHAPTER FOURTEEN

The population of the Aspen Grove region was relatively small, but it included some very fine and interesting people. One such family was the Menzies, whom we first met in 1930. There were five in the family: Mr. and Mrs. Menzies, their son Jim and two daughters, Tibby and Netty. They lived in isolation in a region called Kane Valley and were the only residents in the valley at that time.

The elder Menzies was an extremely well-educated and well-spoken Scot. A man with an iron will, a determined task master, he was noted for his cutting sarcasm.

Mrs. Menzies had come from the Old Country to become his bride, knowing nothing of where she was going. Having experienced only the life of a Scottish society lady, she was completely unable to cope with the rugged life and isolation of Kane Valley. She had accordingly resigned herself to a secluded, lonely life and when we met her, had not been beyond the confines of her own home for many years.

Mr. Menzies had originally pre-empted land in the valley, started a ranch, and set up a home-operated sawmill, with which he cut lumber to order for local trade. One of the girls, Tibby, operated the steam engine which ran the mill, her sister Netty operated the head saw, while Jim and his father did the heavier work involved in the operation. When the mill was working, it was quite something to watch. The entire mill shivered and shook as though ready to fall apart. A cloud of glowing sparks and embers belched from the stack and drifted in whatever direction the wind dictated. Because of this, the mill could run only on rainy days or during the winter months. When not cutting lumber, which was most of the time, they tended their cattle and farm crops. The two girls spent many hours in our camp and frequently arrived with a spare horse for Vi, who would accompany them across the range to check on their cattle.

Since that time, 50 years have passed. The parents have long since passed away and the old mill has fallen apart; but the three maintain their ranch and, despite the passing years, little has changed in their lives. Each one has acquired the allowable acreage of pre-empted land. Their combined holdings now amount to better than a thousand acres. On this they raise several hundred head of cattle. We never pass that way without a detour into Kane Valley to visit our old friends, the Menzies.

Their lives have been closely woven with that of the valley and while they have seldom been away, (Jim was in the Navy during the war), Kane Valley offers them everything they desire. For them to attempt to change at this stage of their lives would be, they agree, a disastrous mistake. They are happy where they are. The world owes them nothing, nor does it offer them anything they cannot find in their own environment. Their inner contentment makes them today a unique family, a privilege to know.

Roundup time was very special throughout this cattle country. One year, the Douglas Lake Company and the Guichon Ranch combined forces for a joint roundup of cattle from the summer range land. The cattle were brought in either to ship and sell, or feed in their winter quarters. Such a roundup called for a party at one of the ranches. Vi and I had been invited to other such occasions and this party was no exception. The majority who attended were part Indian and everyone involved in the cattle drive was there, whether invited or not. The usual stimulant was gallon jugs of logana wine but since riders for both companies earned only one dollar per day in addition to which they had to supply their own horse, few could afford even that, the cheapest alcoholic beverage on the market. The gallon bottle passed around the room hand to hand, each person taking a swig and passing it on—certainly not the most sanitary practice! We followed suit, Vi insisting it was the friendly gesture that counted.

Most of these parties were relatively orderly and good fun for all attending. It was mostly square dancing, and the expertise of both dancers and callers left Vi and me feeling very inadequate. Music was usually provided by a fiddle, guitar and accordion, and the calibre of the musicians astounded me.

On one of these occasions, only the guitar player showed up, much to the disappointment of all the assembled guests. After much to-do in trying to locate a fiddler or accordion player, without success, I suddenly remembered I had a harmonica in the car. I offered to fill in with this, which was better than nothing. Suddenly, the party was brought back to life and so, with mouth organ and guitar for music, the dancing commenced. The two of us played—and played—and played—every square dance tune we could think of. The dance was going very well, and the harmonica seemed to be holding up. Unfortunately, the wooden dividers between each note began to swell from moisture of my mouth. They acted like the teeth of a saw across my lips. Periodically, I had to take my jack-knife and whittle smooth these wooden dividers as my mouth got progressively more painful. Apart from an occasional stop for a cup of coffee, we played until four in the morning. By then, my mouth organ and I were both blown out! I had to stop. To cap the evening, the cowboys, impoverished as they were, took up a collection and offered to pay me for playing, an offer I appreciated but refused. It was quite an evening, one I would hate to have missed.

Not all parties went as planned, for frequently something unexpected happened to add interest and novelty to the evening. At one such event, about midnight, a giant-sized Indian arrived, uninvited and drunk. His mission was apparently to settle a score with someone at the party. Just at the moment when large pots of hot coffee and plates of sandwiches (always made of two full slices of bread) were being passed around, the troublemaker arrived and a fight broke out. Some intervened to try to calm the disturbance. Some joined the fight to assist others. Some seemed to participate simply for the sake of a scuffle. In any case, plates of sandwiches were soon on the floor along with pots of spilled coffee. The fighters slipped, slid and fell in the mess on the floor. Vi and I were interested solely in keeping away from flying fists and falling bodies. We became separated in some way, and a local lad took charge of Vi's protection by guiding her into a pantry, out of harm's way. I became isolated in one corner of the kitchen, where I stayed, relatively safe, until the fracas died down. Eventually the peacemakers overpowered the troublemakers, carried them bodily through the kitchen and tossed them out the back door over the porch railing, into the darkness where they landed in a small ravine. What happened to them after that, I never learned, but at least they did not come back.

In due time, the floor was cleared, more coffee was made and the party resumed. The host of the memorable occasion never ceased to apologize to us for this ignominious party.

Each year with the passage of summer, the nature of my work changed. Studies of bark beetles were completed for the season, and my attention turned to an over-all survey of the intensity of attack which had taken place on the pines during the summer. This was always enjoyable work as it took me over a wide expanse of beautiful country during the best part of the year. The days were bright, sunny and warm, the evenings sharp and crisp, and all the flies and mosquitoes gone for the year. Myriad small lakes scattered throughout that country offered exceptionally fine duck hunting, if one was so inclined. During the autumn months, it became our custom to have either duck or grouse for supper each day. Occasionally, Vi would can one or the other for use at a future date. A novel method of keeping birds for future use was taught to Vi by a local Indian. The bird is first cooked, after which it is packed in an earthenware crock and covered with hot melted fat. The process is repeated as others are obtained until the crock is filled. When desired, the bird is heated in the oven, the fat melts away and the meat is as though freshly roasted. Whether this was only a local custom or an old recipe, we never knew, but it was a very satisfactory way of preserving birds for winter use.

During the second summer, we were joined by another research team doing experimental work on the control of grasshoppers on range land. This work was conducted by Professor George Spencer of the University of British Columbia, with one assistant. While there was no involvement of my work with theirs, we enjoyed their company during the summer. It meant a good deal of extra work for Vi, as we shared the common dining room and meals. Although she worked hard like the rest of us, there was never a suggestion of pay for the work she did.

Among other activities of the grasshopper crew was the development of poisonous baits for use against the pest. Two ingredients for this bait were molasses and arsenic. Storage and mixing of these baits was carried out in an old cabin, where things could be safely locked up. The molasses was contained in an open barrel from which it was ladled. The arsenic was also in liquid form. The two ingredients were measured using the same dipper. The fact that some arsenic might get into the molasses was of no great concern, since the two would be mixed anyway when used. When the summer's work was over, there still remained a good portion of molasses in the barrel. Upon departure, the crew posted "poison" signs within the building and outside, the windows were boarded up and the door padlocked.

Living in the community was a character known to us only as "Three-Finger Slim," locally renowned as a distiller of homemade whiskey. During the ensuing winter, Three-Finger Slim had been boasting about the fine rum he was in process of producing, although I knew of no one who had sampled it. In any case, when the grasshopper crew returned in the spring, they found their carefully locked up cabin broken open and the molasses gone, despite the warning "poison" signs. As for "Three-Finger Slim," he had died during the winter of unknown causes!

*　　*　　*

Like all great outbreaks of native insects, the bark beetle infestation slowly disappeared as the population subsided due to several factors, not the least of which was the absence of any further old growth, mature pine. Thus, after four years of field research, the great bark beetle outbreak came to an end. It had persisted since the early 1920's and, despite the best efforts at control by the provincial government, it ran its course over some fourteen or fifteen years. To date, no plan of applied control has achieved any lasting good as far as this insect is concerned. It is a problem with which we must live and the key to co-existence with it lies in the field of management. Through research in both Canada and the United States, much has been learned and while it will continue to be a problem as long as there are old growth trees in which it can live, losses can be greatly reduced through increasing knowledge of the insect and its host trees.

CHAPTER FIFTEEN

With the termination of my bark beetle project, I decided to return to university for graduate study. I chose McGill University, Montreal, and obtained admission to the graduate school. With rather meagre savings from the previous years, we departed for the East, September 18, 1933.

The graduate school was at MacDonald College, some 20 miles out of Montreal. A more beautiful campus would be hard to imagine. For the following two winters, we lived as best we could on our limited funds. Vi found that by carefully buying and planning, she could provide a well-balanced diet for a monthly total of $15.00. Other students with whom we associated, including several married couples, were in the same financial straits so we adjusted our lives and activities accordingly.

This was at the bottom of the great depression. There were no government grants to assist students, regardless of circumstances. There were no leaves-of-absence with part pay. There was not even any assurance your job would still be there even if you returned with more degrees. It was a time when a person got only what he worked and saved for, and very often not even that. But the classes were full of self-supporting students who managed to get by despite the depression, the lack of jobs, and the complete absence of government handouts.

During my two years of graduate study, Vi worked part-time nursing, a position she obtained through her friendship with a neighbouring graduate nurse.

*　　*　　*

When I returned to my job in British Columbia, my work was located in the Kootenay country, in the southeastern part of the province. The Kootenay, like the Okanagan, is beautiful country, with lakes, mountains and fertile valleys. I was specifically concerned with the portion of the Banff National Park located between Radium Hot Springs and Lake Louise. A large part of this country was forested with pure lodgepole pine, even-aged and all in a mature condition. It provided a prime target for bark beetles, who prefer old growth trees weakened with age.

These forests were undergoing attacks by this insect, resulting in large areas of dead and dying timber. For three summers I witnessed the spread of this infestation throughout the valley, an outbreak, which eventually swept through the entire stand of lodgepole pine, killing most of the susceptible trees and leaving a forest of dead trees and snags. From the standpoint of the aesthetics of the park and the increased fire hazard, this outbreak was calamitous. A forest, however, is very resiliant and, barring fire, will replace itself with new growth in a relatively short time. In this instance, the pine was replaced with spruce and balsam, the natural sequence of forest evolution in that part of the world. It can be assumed the region was primarily a spruce forest which was probably eliminated by fire about a hundred years ago. This was replaced by lodgepole pine, the normal follow-up to fire in central British Columbia. Spruce, however, is the climax forest type in that region. (A climax species is defined as "one that continues to renew itself as long as its habitat is left undisturbed.") The pine served only as a nurse crop until the spruce could again be established. With the maturity of the pine, the young spruce forest had become well established throughout the country. The time had come in nature's calendar for the removal of the old growth pine, which was achieved in a rapid, thorough manner with her very effective tool, the bark beetle.

Nearly 50 years has passed since then. Today the last trace of the original pine is gone, some having blown down, some rotted and fallen, some having been removed by the Parks authorities. Nothing of the original stand remains to indicate the wide-spread kill. In its place there is today a new forest of spruce, balsam and young growth pine, vigorous and healthy, a far superior forest to the original. Few travellers who motor between Radium Hot Springs and Banff are aware of the succession of events that have taken place over the preceding years to result in the forest we now behold.

The entire sequence is an eloquent reminder of the resiliance of nature, when left to her own devices. Man may destroy the forest, pollute the rivers and even tear open the land in his quest for coal and minerals. Nature may do likewise through drought, fire, floods and other natural mishaps. But with the infinite patience of nature, there will eventually be a re-birth of plant and animal life much as existed before the upset occurred. Each spring, the landscape will once again become a sea of verdure, a living environment—forever green.

Nature's progress as measured by man's requirements, however, is slow and unpredictable. Bark beetles today continue as one of the great destroyers of pine, fir and spruce throughout the interior forests of British Columbia. Losses in some regions exceed the allowable annual cut. There is no evidence of a lessening of beetle activity. It will continue in greater or lesser intensity, dictated probably by prevailing weather patterns, so long as mature and decadent timber exists. Demands on these forests require that such devastated areas be cleared and replanted as soon as possible wherever the site justifies the cost, as measured by anticipated returns on the investment.

CHAPTER SIXTEEN

It was in the National Park, while doing field research on the large bark beetle outbreak, I had an unusual and never-to-be-forgotten encounter with a bear.

Vi and I, together with a student assistant, Ken Graham, had just arrived in the National Park in preparation for several weeks of work and had selected the McLeod Meadow campground as a convenient location from which to work. Since our stay there would be for a considerable period, we had brought along a good supply of food, including a full side of bacon. Having selected a suitable campsite, we pitched the tents and stored the food in one of them. While Ken and I prepared the camp, Vi busied herself cooking supper. After supper, we were enjoying a final cup of coffee by the glowing fire in the approaching dusk when, to our surprise, a series of grunts came from the rear of the tent.

"Listen," Vi said, with a look of surprise on her face, "sounds like an old pig." She had no sooner spoken, than a large black bear appeared around the corner of the tent. Now it is a well-known fact that no bear can resist the aroma of bacon. Nose to the ground, sniffing and grunting with each step, he continued, hoping to find an opening into the tent and access to that delicious-smelling bacon.

We yelled at him, but he didn't look up. I grabbed a stick and slapped the tent with a loud sharp crack, hoping this might distract him, but on he came, each step bringing him closer to the open flaps at the front of the tent. Once in the tent, of course, nothing could save the situation. Our food would be spilt or eaten, sleeping bags and clothing would be torn to shreds and the tent would probably be wrecked, for it was not likely he would bother to leave by the way he entered. I was almost at the point of surrender when Vi rushed to my side with a flashlight. "Here," she said, handing me the light. "Flash this in his eyes and yell at him. You might stop him and then chase him away." (Vi was always very imaginative in such an emergency.)

Taking her suggestion and keeping well beyond his reach, I flashed the light and yelled. For the first time, he paused and looked up at me. I continued with the light and the yelling as he stood motionless. His eyes became bloodshot, and from his throat came a deep, gutteral, choking sound. He was beginning to look a bit irritated and I backed off, warily. Returning to Vi, I handed her the flashlight with the warning, "I think he's getting mad. We'd better play it cool."

Fortunately, the old bear got the message he was not welcome in our camp, and ambled off into the woods.

"Well," said Ken, "Now what? He'll surely be back sooner or later."

"I know just how to settle it," Vi announced. "Put all the food in the car and leave a window open slightly. Then, if he returns, we will not be molested in the tent and the food will be perfectly safe in the car." So that was the arrangement. We went off to bed. All went well through the night. Just at the break of dawn, I was awakened by Vi shaking my shoulder.

"I think the bear's back," she whispered. I sat up in bed and listened. "Sure enough," I answered, "he's out there all right."

We peered through the tent flaps. There he was, defiant as ever, and bent on finding a way into the car. First, he put his nose to the opening in the window and got the tempting aroma of food and bacon. Circling the car, he hopped onto the hood, looked in through the windshield, then climbed on top of the car, jumped down from the rear to the ground, repeating this performance several times. Getting nowhere, he finally returned to the engine hood, peered in again through the windshield, seized the wiper in his teeth, spat it out, and then climbed onto the top, where he sat, apparently formulating further plans, while we worried.

Our 1934 Ford V8 had a fabric panel in the centre of the top. Between the fabric top and the inside lining were several inches of fibrous padding, the electrical wiring and a full panel of wire mesh which served as a radio antenna.

The bear began to feel around at the soft texture, and then to circle in the manner of a dog preparing himself a bed in the flower garden. It was now quite evident he was planning to tear open the top of the car. Losing the food was one thing. Having the inside of my car wrecked by a bear was quite something else. The car was mine, not the government's! The situation had taken a very critical turn. We were desperate.

Vi threw cans of milk at him, items we had not considered necessary to lock up. They had no effect. The cans bounced off him like falling leaves. I managed to find a long pole with which I hoped to spear him and drive him off, but he simply swung at the pole each time it came near him and knocked it away like a straw.

Ignoring us, he finally sat on his haunches and with both front paws, proceeded to claw and rip through the top. Next came the padding, the electrical wiring and then his claws took hold of the wire mesh. All that was left of the top came away, leaving only the cross ribs of the roof. He was as good as in the car. He merely had to slip between the cross ribs. The bear lowered his head into the car, then put his paws onto the back of the front seat. With his head and shoulders over the steering column, only his black rump was exposed on the top of the car. It would be mere seconds before bruin would be inside the car and we would be helpless. I began to visualize my car torn to shreds inside and to wonder why we had not let him in the tent in the first place.

In a moment of frustration and desperation, I grabbed the front door of the car, opened it and yelled directly into the bear's ear. It came as a shock to the bear. It shocked me too, when I realized how close I was to that bear's head! Startled, he wiggled back onto the top of the car. I had not anticipated this reaction, but now that he was out, I saw my chance. I jumped in. If only I could get the car moving, I might roll him off. Luckily, I had left the keys in the ignition, a thing I seldom do. As the front seat was piled with food, it was necessary for me to assume a partially standing position, leaning over the steering wheel. With the bear still on the top, I started the motor and pressed the accelerator to the floor.

While all this was going on, the old bear was also thinking. Realizing I was blocking his way, his most immediate job was to dispose of me. He returned to the opening, lowered his head into the car and with his nose only inches from the back of my head, reached in and took a swing at me with his paw. Vi, standing at the side, quite helpless to do anything, witnessed the whole thing and as the bear struck, let out an ear-splitting scream of horror.

Simultaneously, I released the clutch and with the accelerator floored, the car lunged forward, throwing the bear backward and causing him to miss my head. He was once more back on the top of the car. From there he slid and rolled to the ground.

So intent was I on all these happenings, I forgot the car was moving and had to be steered. Just as the bear rolled off the top, the car came to a crashing stop. Vi assumed I had been hurt by the bear and came rushing to me in a state of alarm. "Are you badly hurt?" she called, extremely concerned.

"No," I said, "I'm not hurt. Why?"

"Well," she said looking at the front of the car, "why did you run into that tree?"

I had ploughed into a large poplar tree, crushed the right front fender and headlight, broken the bumper and buckled the engine hood. Distracted by the bear and what was happening on top of the car, I had no recollection whatever of hitting the tree, or even of stopping. Meanwhile, the bear had frisked across the meadow, bouncing like a fawn at play, and disappeared into the woods.

The car was now a sorry looking sight; top gone, inside lining hanging in shreds, front crushed in. But at least it still ran.

My next step was to report the incident to the Parks Headquarters, since it had occurred within the bounds of the federal park. Their main concern was that the bear represented a hazard to tourists and as he now knew how to get into cars, where children were frequently sleeping, he must be destroyed.

For the next three nights, a warden was posted at the camp, but the bear failed to appear. Nearby was a Parks telephone line, draped from tree to tree, for use in case of fires. I was given a key to the phone and told that should the bear return, I was to phone the warden who would hurry over to the scene and shoot the bear.

A week passed with no sign of the marauder. Sunday morning, we had just finished our breadfast of bacon and hot cakes when the bear sauntered out of the woods. Without hesitation, he came directly to our camp. Leaving Vi and Ken to entertain the bear, I dashed to the telephone line and called the warden, who promised to get out within half an hour. In the meantime, Vi and Ken held the bear's attention by pouring over and under logs all the syrup we had. He immediately licked it up.

Returning out of breath, I found the bear had progressed to where we had been eating and had licked clean each pot, pan and plate. He seized a bar of soap which he immediately spat out and sent flying through the air with one swat of his paw. He tried the still hot frying pan, which, after it burnt his tongue, quickly followed the soap.

He next turned his attention to the tent. It was now a matter of holding his interest until the arrival of the warden. Vi undertook this mission by cutting pieces of bacon and tossing them to the bear. As he devoured each piece, he took another step towards the tent. The slab of bacon grew smaller and smaller as the bear came closer and closer. Just as the bear arrived at the tent opening, the warden roared into camp, skidding to a stop a few feet from the bear. As the warden leapt from the truck, rifle in hand, the bear, realizing he was in danger, disappeared around the tent and scurried into the woods.

The disappearance of the bear did not end the story for me, however. The question uppermost in my mind was: who is responsible for my car's repairs?

The Parks disclaimed any responsibility since I was not working for them, even though the work was a joint undertaking with the Department of Agriculture. My department said they would look into it, but knowing how long that might take, I approached my insurance company, claiming the damage should be covered under the theft clause, since the bear was definitely a thief and my policy failed to define exactly what was or was not a thief! The insurance company replied, "We have never had a more interesting claim than your story of the bear and your car. As a matter of fact practically everyone in the block has read your letter. It is the considered opinion of all that, had we been in your place, we would have been hightailing it down the road, leaving the car to the bear. In any case, we regret this can hardly qualify for damages under the theft clause of your policy."

The company might well have paid the claim to avoid the publicity which would have resulted from a court case, but in the meantime, my employer, the Department of Agriculture, advised me to have repairs completed at their expense.

Eighteen years later, in 1954, the story was related to the Canadian Parliament, when the Minister of Agriculture read to the House of Commons my report of the incident. A Montreal newspaper report of this episode read:

"The memory of the bear was honoured in the Commons when the Minister of Agriculture, Jimmie Gardiner remembered him when the House of Commons was discussing the liability on cars driven by government employees. The Honorable Gardiner asked the house to leave the question with him until the next day when he would read to them a case history which he thought would answer the question."

The following day, armed with my account as I had reported it to my superiors in Ottawa at the time of the incident, the Minister read them my bear story. Thus it went on official record in the House of Commons. The verbatim account from Hansard ends with this comment:

"MR. PEARKES: That's a good fish story."

I like to think that through the succeeding years, a grandfather bear roaming the woods in British Columbia would gather a company of starry-eyed cubs about him and tell them how he once went for a ride on the top of a car, and had his name mentioned in the Canadian House of Commons 18 years later!

Never did a bear attain such fame and distinction.

CHAPTER SEVENTEEN

The year 1937 stands out as monumental for me. Just as I was preparing for another season in the Kootenay country, word came from Ottawa that I was being transferred to Winnipeg to start a new forest insect research centre for the forest regions of central Canada. It was to be situated on the campus of the University of Manitoba.

A new Chief of the Forest Insect Division of the Department of Agriculture had recently been appointed. He was Mr. J. J. deGryse, one of the most learned and brilliant men with whom I have ever been associated. When he assumed his new responsibilities, he immediately initiated a Canada-wide forest insect survey. The survey was to have many objectives including the determination and recognition of our major forest insect enemies, their biology and potential as destroyers of forests and forest products; their distribution, the relationship of one species to another; factors of natural control; their place in the general ecology of the forest; appraisals of damage resulting from insect outbreaks and so on.

It was an imaginative and challenging programme, one that would involve not only federally employed personnel, but all provincial forest service personnel and forestry personnel of major forest companies as well. Since that beginning, the programme has undergone numerous changes in techniques and operation. Biologically and economically it has become one of the most important and fundamental aspects of the Canadian Forestry Service. The whole concept has been adapted by co-workers in the United States and is now an undertaking of continental importance.

The organization of this survey throughout northwestern Ontario, Manitoba and northern Saskatchewan, together with the development of various research projects in those forested regions, constituted my responsibilities while in charge of the Winnipeg laboratory.

The transfer of personnel by government during the depression years was vastly different from what it would be today. I was told only my personal travel expenses would be paid; there would be no reimbursement for transfer of my personal effects. Unless I was prepared to pay the freight costs myself, they advised me to sell or otherwise dispose of any furniture or other possessions. There would be no travel or hotel allowance for my wife. Since Winnipeg would be my new headquarters, I would be entirely on my own financially once I set foot in the city. Furthermore, as my transfer was immediate, Vi was left to sell all our things, including the car, which she subsequently did by public auction. In those depression days, one hung on to a job in spite of what one might consider to be somewhat unreasonable demands from Ottawa. The government, of course, like the people it represented, was in deep financial trouble.

My territory in this new job was comprised of the forest regions of central Canada, an area extending from Lake Nipigon in northwestern Ontario, westward through Manitoba, northern Saskatchewan and into Alberta, and including three time zones. My salary was $135.00 per month, less a ten percent cut due to depression budgets. Nevertheless, I was glad to have a permanent job, which few had, and pleased to have been selected to head up this work. It was all very exciting and challenging.

I expected that leaving the mountains and forests of British Columbia for an assignment in the prairie provinces would require a great deal of adjustment. Much to my surprise, it was the beginning of one of the most enjoyable eras of my life. Contrary to my expectations, the prairies hold a charm and fascination that defies explanation.

I once wondered how prairie visitors to British Columbia could dislike the mountains and experience a feeling of claustrophobia when among them. I was astonished when a prairie visitor once said to me, "If I could only push these mountains aside so I could see space." Such a feeling is easily appreciated after living in central Canada.

Instead of mountains, I found myself coping with lakes, thousands of them! Small ones, big ones, all shapes and sizes were dotted over the map in the Pre-Cambrian Shield of Ontario and through the forest area of eastern and northern Manitoba and Saskatchewan. If I ever wanted space in which to move, I had it here. For the next eight years, my life and work centred about these forests and some of their problems.

One thing that soon became apparent was the difference in the mode of life of those who lived and worked in these forest regions. Here, the canoe was an essential means of travel. Never having lived where the canoe was anything more than a recreational item, I had much to master regarding its use. I had also to learn about black flies, for while I thought I knew about them, I discovered I had never lived where they were such a diabolical problem. To a newcomer, those insects can be not only an annoyance but also decidedly poisonous. On a sensitive person, they can cause such swellings as to completely close the eyes and almost totally incapacitate a person. Fortunately, one soon develops an immunity and learns to live with them as well as with the mosquitoes. Luckily for me, I suffered less than many other newcomers with whom I was acquainted, since I am not very sensitive to such irritations.

At the time of my introduction to this country, these forests were relatively untouched. They varied in their ecological make-up and forest growth, disappearing entirely as they merged with the grass lands of the prairies or the tundras of the north. Roads were few, none were paved and the tourist business was in its infancy.

It is doubtful if any place offers a better habitat for wild ducks and other waterfowl than some of the lakes of northwestern Ontario and eastern Manitoba. Rice abounds in these lakes. Wild rice is a delicacy for both men and waterfowl, and commands an exorbitant price on the market, if and when available.

During my working days through that country, the government exerted rigid control over the management of wild rice and its harvesting was restricted to Indians.

Wild rice grows in relatively shallow water, reaching a height of two to six feet above water levels. By the rather primitive method of harvesting developed by the Indian, he paddled his canoe through the dense growth and with a long pole swept it over the gunwale of the canoe. While he held it in position, a second person beat the grain from the plant with a flail. The grain thus dislodged fell to the bottom of the canoe. When he had sufficient load, he returned to his camp. Here the women and children swept the rice from the bottom of the canoe and placed it in an open tub.

At this stage, the rice grain was enclosed in a fibrous sheath. It was then carefully heated over an open fire until the sheath opened, releasing the grain. Then shaking of the tub caused the heavier grain to settle on the bottom; the lighter chaff was scooped off the top. Final sifting was accomplished simply by pouring the rice from a height on a windy day when the fine particles were blown away, leaving only clean grain. The final product was sold through the Federal Department of Indian Affairs. Only broken or culled grain could be sold locally by the Indians.

While these methods were a bit crude and primitive, they suited the Indian very well. It was a big event each autumn when the whole family set up camp on the shore of one of the lakes and proceeded to reap a harvest nature had provided free of charge.

Today, wild rice production has become an industry of considerable importance and the harvesting, processing and marketing has increased greatly in Manitoba. The young native pickers use a variety of mechanical harvesters while the elder native people still harvest by canoe and the hand flail method. Some 50 percent of the rice produced in Manitoba is now harvested by non-Indian people. Nearly all the freshly harvested rice is sold green at lake side and flows through buyers and processors in Ontario and the U.S., from which point it is processed, packed and marketed.

I had my first experience with wild rice when I attempted to navigate, by canoe, a lake in which there was a bumper crop. I was travelling with a companion, Ray Lejeune. The lake was solid with rice plants from shore to shore and while it had seemed navigable at the start, it became increasingly dense as we progressed. By the time we reached the centre of the lake, the rice plants towered above our heads to form an almost impenetrable jungle. The shoreline was invisible so it was impossible to see how far we had to go to reach the other side of the lake. The canoe could no longer be paddled but had to be pried through the dense growth with great difficulty. The water was much too deep to wade so we pushed and pried, hoping all the while that neither paddle would break. It was late afternoon before we reached our destination where we were glad to camp for the night.

The following year I was travelling the same route, expecting a repeat of the same difficulties but to my amazement, there was no rice. This was a year of abnormally high water levels; too deep for the growth of the plant. Only in years of optimum water levels does the Indian enjoy a good harvest.

CHAPTER EIGHTEEN

On a remote shoreline of the eastern side of Lake Winnipeg stood a structure incongruously impressive for such an isolated setting. Its rustic design indicated it had been built to blend with the wild surroundings, while its size indicated the builder's optimism regarding its future as a resort. Judging by the lack of clientele, however, it appeared to be a resort in name only. It was run by a lone woman who seemed to have lived too long in isolation.

I was flying over the forest regions east of Lake Winnipeg in the course of mapping the extent of the jack pine budworm infestation when the pilot, Huey Smith, turned to me and pointing at the lakeside edifice said, "This is where I figured we could stop for the night. The Forest Service has some gas stored here." Huey had stayed at the "resort" before and knew the woman who ran it. "Don't expect much in the way of service," he warned. Huey circled the place to give me a good view from all sides and then, after skimming along the water to a stop, we tied the plane to the float and walked up to the building. The proprietress was sitting on the steps of the verandah, stray hair straggling over her face and a cigarette dangling from her lips. Without removing the cigarette or changing her position, she greeted Huey with, "Well, you're back. What do you want?" In response to this enthusiastic reception, Huey told her he wished to stay the night.

"In that case," she replied, still sitting, "You can use the cabin you had before. You know where it is. Now I suppose both of you want your suppers." We assured her we did. "Well, you can get me in some wood, and then light the fireplace while I fix you a meal."

She was a very small woman, probably less than five feet tall, and as slight as she was short. Her diction was precise and she spoke with a very cultured English accent. In addition to her refined speech, she had an impressive range of profanity.

That evening we sat with her before the glowing fireplace. She perched herself on a high stool, her legs wrapped around its spindles, a cigarette hanging from her mouth. Ashes dropped onto her blouse as the cigarette flipped up and down when she spoke, and never once did she make an effort to brush them off. No sooner was one cigarette finished than a new one was lit and remained in her mouth until practically burning her lips. She referred to everyone who had been there before or with whom she had any business dealings whatsoever as either a "son-of-a-bitch," or a "bastard." We, ourselves, seemed to pass in her judgment as normal humans, at least while we were there, despite the fact we worked for that "god-damned government."

When it came time for bed, Huey and I retired to the cabin she had assigned us. It was quite clean and presentable with twin beds. Looking at his bed, Huey remarked, "You know, I was here a month ago and I had this same cabin. I wondered then if she ever changed the bedding." With that, he proceeded to turn back the sheets. "Just to check, I marked the lower corner of one of the sheets with an ink mark."

We looked and there was the tell-tale mark! How long the bed was used without a change of sheets was anyone's guess. We figured it was quite a while. We took both beds apart, reversed the sheets so that what was underside was now uppermost and assumed we were sleeping on the clean sides. In any case, we slept without trouble.

With morning, we returned to the lodge to see about breakfast. Our hostess was up, sitting in her kitchen, a cup of tea on her lap and the inevitable cigarette in her mouth.

"I suppose you boys want your breakfast now," she remarked. That was the general idea, we advised her, whereupon she replied, "There's the bacon and eggs over there, and if you make up that stove, you can go ahead and cook yourselves something."

We washed some dirty dishes from the night before and emptied the dishwater into the sink. It disappeared with a gurgle and a sucking sound, only to reappear beside a path at the back of the house. I couldn't help wondering how clean her kitchen was as well as the food she served. However, nothing could be done about it, so I just enjoyed myself and imagined that what I couldn't see was sanitary.

We cooked and ate our breakfast while she expostulated on her troubles with the government and the people with whom she had done business in the past. Breakfast finished, we packed our gear, paid her for the lodging and meals and left her as we had found her, sitting on the steps of the porch, cigarette in her mouth, ashes falling down her blouse—a rather pathetic little figure, in her wilderness retreat—alone.

CHAPTER NINETEEN

Autumn was approaching. I was flying over northern Manitoba at the end of a successful year of co-operative work between the Manitoba Forest Service and the federal forest insect survey, which had included the introduction of some beneficial parasites for use against the budworm. This was our last trip through that country before winter. It was mid-afternoon. The weather had been threatening all day. Heavy clouds hung low on the western horizon. We were flying west over a large expanse of uninhabited country. With me was the District Forester, Bob Harvey from The Pas, my co-worker Ray Lejeune and the pilot, Bob Parkin.

Suddenly the landscape disappeared behind a curtain of grey. "Snow ahead," the pilot called. "We'll see what it's like, but we'll probably have to change our plans."

Within minutes we were flying blind in heavy snow. We turned southward, but again the storm cut off our route.

"There's only one place left to go," Bob said, as he made another turn—"Herb Lake. We can get down there and stay the night." We changed our course and, after some time, a lake appeared. As we circled for a landing, I had my first glimpse of the community of Herb Lake. It wasn't much to see, just some unpainted shacks nestled together at one end of the lake. The railway running between The Pas and Churchill crossed at the other end. Why the community was not established on the rail line no one seemed to know. Residents figured someone had settled there before the railroad went through and the community remained where it was originally located, despite the fact all supplies had to be transported down the lake by boat.

The "hotel" consisted of four unpainted huts which had been dragged together like four spokes on a wheel, with a connecting room as the hub. In this central area was a stove that provided heat for the entire hotel. Of the four rooms, three served as bedrooms and the fourth was a beer parlour. Ray and I shared a double room. The proprietor's daughter, Dulcie, proceeded to instruct Ray and me about the operation of the establishment.

"If you want heat," she advised, "go down that path at the back of the hotel. You'll come to a gate. Go through that gate and soon you'll come to a wood pile and an axe. Split all the wood you want and bring it back and light the stove. About the water, well, just tell Bob Parkin to get it. He's been here before and knows all about that." With these instructions, Dulcie disappeared.

Ray and I followed Dulcie's directions—down the path, through the gate, find the woodpile, the axe, split two armloads of wood, return to the hotel, light the fire and get some warmth in the establishment. At this moment Bob Parkin appeared, and we gave him his assigned job of getting water. "Dulcie said you know all about it," we told him. "Damn right I know—you take those two pails down to the lake and bail the water from the end of the dock," he replied in disgust as he disappeared, taking the pails with him. We made the supply last for the duration of our stay!

As the day was still young, and there was nothing else to do, we decided to give the local pub some business. The only other customers in the place were some old-time prospectors huddled around a table in one corner of the room. No sooner were we seated than the proprietor plunked four bottles of beer on our table, and commented "Compliments of the house," and vanished.

The owner was a slightly-built man, an English Jew who spoke with a very cultivated English accent, and carried himself with grace and dignity. We were very surprised at his generosity, especially when he appeared again and placed before us another four bottles. When we had finished our first bottle and were into our second, he arrived with a third bottle for each, remarking only, "Enjoy yourselves, gentlemen," before disappearing.

Although I knew the first bottle was free, I assumed the following rounds had been ordered by the other three of our party, although I had not noticed anyone paying. Certainly I could not consume a fourth bottle, so I decided that after supper I would do my share of buying for the others. When we discussed this among ourselves, it turned out each thought the same thing. None of us had ordered or paid for anything. Obviously, the hosts were the old prospectors huddled around the table in the corner of the room. We agreed that after we ate, we would return to the pub and each of us would buy a round for the house, although we ourselved had consumed about as much as we wanted for one day.

The only eating establishment was in a log building about a half mile down a road through the bush. It was here that all the hotel guests, when there were any, were sent for a meal. Although inconvenient, it was well worth it, for the owner of the establishment apparently got his main pleasure from serving outstanding food and receiving compliments from his patrons.

Supper finished, we returned to our hotel and were greeted outside by one of the local prospectors with a case of beer. "For you, my friends," he said, taking a couple of fast steps to keep from falling over. We thanked him but suggested he bring his beer into the pub where we all could share it. He readily agreed to return to the pub but insisted the beer was ours and should not be shared.

As arranged, we returned to the pub. Each of us ordered a round for the house but in so doing, we simply ignited an explosion. From that moment on, beer arrived at our table as fast as the proprietor-waiter could deliver it. Open bottles covered our table until it could hold no more. Still he brought beer which then had to be placed on the floor. We told him to stop bringing beer but he informed us the fellows at the other table were doing all the buying. We invited them to join us to help consume the beer they had purchased. For the next hour, we heard astounding tales of great gold mines just waiting to be discovered, stories of lost mines, lost dreams, missed opportunities and visions of wealth just waiting to be found.

The evening at the pub came to a sudden ending with the appearance of the proprietor's daughter, Dulcie, who proceeded to take away the four Coleman lanterns which lit the place. Her parting comment was, "Sorry to take the lights. We need them for the bingo game."

With that, the pub, its patrons and the many opened bottles of beer were left in total darkness. Of course, it is just possible that the very accommodating proprietor knew full well what was going to happen and therefore encouraged a rapid sale of beer

before the lights went out. In any case, it was our opportunity to
excuse ourselves, leaving our new-found friends in the darkness
with countless bottles of beer, while we spent the balance of the
evening at the bingo game.

*   *   *

The weather cleared, we departed the next morning, visited
several areas of timber, and arrived at the end of the day at Cross
Lake. Here we planned to spend the night at a Catholic Indian
mission. Cross Lake was completely isolated from the outside
world, accessible only by air or water.

Flying in for a landing, I was admiring the picturesque setting
of the lake, when a two storey, stone and concrete building
seemed to appear from nowhere. This was an Indian school
serving the native population throughout the wilderness north of
Manitoba. In charge was Father Trudeau, a well-known mission-
ary who had served many years among the northern Indians and
Eskimos. Father Trudeau and Brother Minard met us at the dock.
They were old friends of the District Forester and the pilot, but
strangers to Ray and myself.

After preliminary greetings, we were asked to bring our possessions with us to the mission where, we were assured, we would be
most welcome to stay the night. Father Trudeau took us to his
study, where we sat and listened to accounts of some of his dog
team travels in the arctic. Ray and I, and no doubt the other two,
grew hungrier and hungrier, each hoping we might soon be
offered something to eat. Eventually, one of the nuns came into
the study and spoke quietly to Father Trudeau who then asked us
if we were hungry enough to enjoy eating a few boiled fish heads.
The menu didn't sound too appealing but anything was better
than hunger pains. Without waiting for our reply, he led us to the
dining room. There, spread out before us, was a sumptuous meal
of moose steaks (out of season), whipped potatoes, baked carrots,
sliced tomatoes, raspberries, cookies, doughnuts and coffee. The
repast finished, we returned to Father Trudeau's office, played
four games of cribbage and were then shown to our rooms. Ray
and I shared one with private bath and twin beds, each of which
was covered with an exquisitely made patchwork quilt.

124

Following a good night's sleep and an elaborate breakfast, we were taken on tour of the school from the furnace room, through the kitchen, bake shop, laundry, the boys' woodworking room which was equipped with power tools and sewing room with electric sewing machines for the girls, classrooms, and chapel. When I saw all the power tools and modern facilities for instruction, my first question was: how on earth can all these modern things be of any use to the Indians, who live, and will continue to live, in isolation in the northland? Most Indians had no desire to change to a white-man's existence, I felt.

In reply, Father Trudeau said that he agreed with me in regard to the Indian in the present generation. His concern was for the future Indian, yet unborn, who would face a very different north, a country awakening to modern developments of the white man. He realized they could not now make much use of the trades and skills taught at school, but knowing what the school offered, they as future parents would be much more receptive to the idea of their children becoming educated in order to deal with the ever-changing north.

I find it very sad that the Indian must have white man's life style forced upon him whether he likes it or not, and cannot have the life and world to which he was born, although I realize that is impossible in the light of modern society.

After another elaborate meal, we left the hospitable and interesting mission and returned to The Pas.

CHAPTER TWENTY

We lived for eight years in Winnipeg. In addition to being in charge of the new forest insect research laboratory, I was also an honorary lecturer at the University. To be an "honorary" lecturer simply means you do the work without any pay. It was a co-operative gesture arranged by Ottawa, as our lab was on the University grounds. For two winters I gave a graduate course leading to a Master's degree. I finally begged out of it, since it was too much work and interfered with my official responsibilities.

My life in Winnipeg was eventful in more ways than one. First, and most important, my daughter was born. We called her "Donella," a family name through several generations of Richmonds. The start of a family, of course, altered our mode of life considerably, but so did the nature of my work.

Another event of major importance occurred during the first years in Winnipeg. Shortly after our arrival, I was approached by a friend to buy a ticket on the Irish Sweepstakes. I had purchased numerous lottery and raffle tickets before and had promptly forgotten them. This ticket was different. From the time I purchased it, the ticket haunted me. Came the day of the draw in Ireland, there was scarcely a moment the ticket was not on my mind. In mid-afternoon, a telegram arrived for me. I knew it could not be from Ireland and felt silly to have even thought of it. I opened it and sure enough it was *not* from Ireland, but from Ottawa, about some routine business. The day finished, the offices closed and I went home. No sooner had I touched the knob than the door burst open and Vi was standing there in wild excitement. "We've drawn a horse in the Irish Sweepstakes!" she exulted. I could hardly believe my ears.

A telegram had arrived at the office after closing time. The messenger told the janitor I had drawn a horse. The janitor phoned Vi and read her the telegram. A strange way for the contents of a telegram to be handled, but everyone seemed too excited to keep it confidential. The Irish Sweepstakes represented a fortune in those depression years when a dollar was worth as much as $20.00 is today.

Foremost in our minds was to get the ticket which was at the office and to see the original telegram. Without stopping, I returned to the office, unlocked the door, found both telegram and ticket and returned with them to Vi. The telegram read:

"CONGRATULATIONS COUNTERFOIL N V 79217 HAS DRAWN DAVY JONES HOSPITAL GRAND NATIONAL SWEEPSTAKE FOR PRIZES SEE BACK OF TICKET HORSES NOT PLACED FIRST SECOND THIRD 460 POUNDS STERLING SENDING INSTRUCTIONS APRIL 1 PLEASE AWAIT—HOSPITAL TRUST."

I had drawn a horse which could be worth a small fortune. The best thing for the moment was to lock the ticket in safekeeping until such time as I received further instruction from Ireland. Accordingly, I phoned a friend of mine, Jim Souter, who was manager of a suburban bank. At his suggestion I met him at the bank that evening and the ticket was safely locked away until needed. For two days, we dreamed of winning the big race; then the bubble burst. Davy Jones was scratched. Nevertheless, there was a prize for us of $2,200.00, which in 1937 was a lot of cash. Sweepstakes were illegal in Canada in those years, so it was arranged that the counterfoil should be sent to Ireland in the bank's mail and the money would be deposited to my credit in a bank in Ireland. I would then arrange a transfer of my account in Ireland to my bank in Winnipeg. In due time, the transaction was completed and with ready cash we were able to negotiate the purchase of a very attractive bungalow on the Red River in Elm Park. The financial boost this lucky win gave us had a profound and lasting effect throughout our lives.

In Winnipeg, Vi and I found a warmth of hospitality and friendship traditional to the prairies, and of the convivial friendliness of the early west. I was already a member of the Kinsmen Club, a service club of business and professional men under the age of 40 and had, therefore, a nucleus of future friends. The Winnipeg club was the largest in Canada and certainly the leading service club in Winnipeg, comprising a membership of 125 men. I eventually became its President and the benefits I received through that experience were beyond measure.

Then war was declared. For the next six years, nothing was normal, except nature's insects. Although there was no letup in work, there was much realignment in activities as they related to the war effort. The production of timber, pulp and especially cellulose rated top priority for there was an ever-increasing and insatiable demand for forest products. Personnel became scarce as all available manpower was absorbed by the armed forces. To prevent total disruption of what Ottawa considered essential work, positions of key personnel were frozen and they were refused release to join the armed forces even if they so desired. My assistant, Ray Lejeune, after qualifying for technical duties in the Air Force and having actually joined, was at the last minute refused release by Ottawa, and forced to return to his former duties, much to his displeasure, but my relief, for we could ill afford to lose him.

One of the principal thrusts of our work was the salvaging of old fire-killed timber in northern Saskatchewan and Manitoba. Here were located thousands of acres of standing timber, killed by previous forest fires. Lacking bark, bleached almost white by weather over the ensuing years, they were destined to blow down eventually, to rot on the ground and return to the forest floor. It was, in fact, an enormous resource of wood for which there had been no market. With the accelerated demand for timber, particularly for the production of cellulose, the government of Saskatchewan promoted the utilization of what had been a wasting resource. There were several factors militating against its acceptance by industry, these being the presence of charred, burned knots; the damage sustained over the years by wood-boring insects and the presence of rot and decay. The practicability of using the timber was uncertain until investigation determined how far such deterioration had progressed. This entailed a lot of field work through the stands in northern Saskatchewan and studies in the mills in The Pas, Kenora, Fort Francis and Thunder Bay. Once the value and use of the timber was established, the province of Saskatchewan discouraged further cutting of green timber until the fire-killed trees had been utilized. Most of these went to the pulp mills in Ontario.

While working in northern Saskatchewan, I had a unique encounter with a family of timber wolves. I was walking along an old unused logging road, through an area that once had been an expansive spruce forest. As a result of past fires, there was now nothing but hundreds of acres of fireweed. There was not so much as a standing snag, stump, or anything to indicate a forest had once clothed the land. My destination lay at the end of a trail which branched off the logging road.

I could tell by the tracks that there was a family of timber wolves ahead of me, two adults and two cubs. I was not especially concerned, since I thought the tracks could have been made the previous day. Presently, however, I came upon two wet spots on the road. As it was a hot afternoon, it was quite evident the animals were not far ahead of me, but since I would soon reach the point where the trail diverged from the road, I figured the wolves would, in all probability, continue along the road and I would be rid of them. I gave the matter only cursory thought.

I reached the point where the trail led away from the road, but much to my surprise, I noticed the wolves, too, had changed course and were heading in the same direction as I.

Now as I walked along the trail, I felt a good deal of apprehension, for no animal can be considered harmless if there is a threat of danger to its young, and these wolves had cubs with them. Few animals in their wild state are dangerous unless convinced that they must protect themselves or their young. Throughout my life, I encountered many, many animals in the wilds, yet never had an occasion to kill one.

The wolves seemed to be keeping well ahead of me. Then, without warning, Daddy wolf leapt from the fireweed onto the trail no more than 50 feet from me. He stood motionless, facing me, and stared. I too stopped, my hair on end. I hoped he might give some indication of retreat, but he did not. With an air of defiance, he moved slightly from side to side, never once taking his eyes from me. I was really scared. I had nothing with which to protect myself except a boy's axe. There was no tree or snag I could climb; nothing but miles of fireweed. If only I had a rifle, I thought. Fear, an animal's best defensive weapon, now seemed to dominate his actions. I shared with the wolf the instinct for self preservation, and accordingly retreated.

Keeping my eyes fixed on him, I backed away very slowly. The wolf seemed content to remain where he was, blocking the trail. I got the message, the trail was closed to further travel. I did not argue the point. I left him standing on the trail while his mate and the cubs waited in the fireweed.

Work on the salvage and utilization of fire-killed timber extended through several years. Concurrently, I was occupied with a serious problem with jack pine budworm in southern Manitoba and northwestern Ontario. The jack pine budworm, a close relative of the notorious spruce budworm of eastern Canada, had created much concern to the forest industry in the two provinces, particularly the pulp mills in Kenora, Fort Francis, Dryden and Thunder Bay.

While the pine budworm worked its destructive way from west to east, there was a steady progression of the eastern spruce budworm from east to west. The two strains came together in the vicinity of Lake Nipigon, which was also the dividing line between the regions of responsibility of the Winnipeg laboratory and the work of its counterpart at Sault Ste. Marie. Accordingly, it was decided to undertake a joint study of these two species of budworm in the region where the populations overlapped. A study centre was established at Lake Nipigon which was jointly directed by Dr. Carl Atwood (father of author Margaret Atwood) and myself.

Just as the work was well in progress, I was called back to Ottawa to assist during a period of serious illness of my chief, Mr. J. J. deGryse. Although I was only supposed to fill in over a short emergency, it was, in fact, the beginning of another chapter in my life.

CHAPTER TWENTY-ONE

I had been in Ottawa only a short time when it was decided I should be transferred there permanently. I was no sooner settled, however, than other alterations were made to my plans. Because of changes proposed in our operations in British Columbia, someone was needed to fill the position of the Officer-in-Charge in that province. Dr. M. L. Prebble, the incumbent, was being moved to administer the work at Sault Ste. Marie. In view of my familiarity with the province, it was decided I should go to Victoria rather than Ottawa. I had no objections to returning to B.C. and to all my old associates.

It was like going home, in some respects, although my early years in B.C. had been restricted to the interior of the province, which is completely different from the coastal region. My work now was in the most highly industrialized forests of Canada, different from anything I had known before. Technically and geographically, it was a new and different world.

Unlike many people, I find the coastal forests much less attractive than the Interior Dry Belt, particularly as a place to work. Though unsurpassed in beauty and a feeling of permanence, the old growth forests of the Pacific are overwhelming, towering above you like huge green curtains ready to drop and smother you. The incessant rain, the damp soft moss underneath, the scarcity of bird life and the heavy undergrowth through which one must struggle, all contribute to a pronounced feeling of claustrophobia. Even on a cloudless day, the sun scarcely penetrates the dense canopy of foliage, and one can work all summer in such timber and return as untanned as if one had worked indoors. The big timber associated with the B.C. coast and the Pacific rain forest is being rapidly depleted in most regions and soon will be found only in areas reserved to perpetuate a sample of the forest that once clothed the Pacific slope.

With the end of the old growth forest, we enter an era of second growth, timber of small diameters by comparison with old growth. In this transition period, it is anticipated there will be a shortage of forest production, in terms of our ever-expanding forest industry. The forecast shortage will persist through the next 40 to 50 years, until our second growth timber comes into full production, which will not be until well into the next century. This impending shortfall is an indication of the urgency for improved management in our techniques of harvesting and utilization and an intensification of forest protection. For the professional forester, these challenges greatly exceed those he would find in working with old growth stands.

During the 1940's, when I first moved into the coastal forests of B.C., there were few roads into the more remote forest regions and travel was restricted to either backpacking and fly camps or to water transport. There were no commercial scheduled flights and travel up and down the coast was mostly by ship.

In order to conduct our work satisfactorily, it became necessary for us to own and operate a vessel suitable to serve both as a field laboratory for the Forest Insect Survey on the B.C. coast, and as living accommodation for those who worked from it.

After much negotiating and investigation, we finally acquired a ship from War Assets, which had a fleet of excellent coastal vessels available. Guided by the experienced personnel of the B.C. Forest Service, we selected the best of these ships before they were put up for public sale. It was a 60-foot halibut-type packer built by Star Shipyards, New Westminster, during the war, as a water transport for the army. To suit our purposes, its superstructure was remodelled to accommodate a total of seven men, with a stateroom for each, a shower room, a compact but efficient laboratory and a galley and storage room. As it carried scientific personnel who were classed as "passengers," it was designated a passenger carrier, as defined by the Federal Department of Transport, and was, therefore, subject to their many legal requirements. It had to be operated by a qualified skipper who held the necessary papers for passenger transport and had to pass periodic engine inspections and to have regular lifeboat drills, to say nothing of its year-round maintenance. It was the first and only boat owned and operated by the Department of Agriculture in

Ottawa, and certainly no one could know less about boats than the people handling the administration of that department. The expense of operating a ship of that size was a real shock to them. For instance, its bank of 16 storage batteries lasted no longer than a battery in a farm tractor or truck. When it became necessary to replace all 16, the cost created a near crisis in Ottawa. The people there found it hard to believe the expenses and problems involved in a ship's operation, and they periodically wrote for advice to the B.C. Forest Service or the Department of Fisheries, both of which operated many vessels. These agencies found some of the letters very amusing, and would phone me and read to me the latest they had received.

Nevertheless, the *J. M. Swaine*, as we named her, served us well and carried her personnel in total safety through some terrible waters, where a vessel of lesser seaworthiness would not have dared to go. It covered the entire coast of B.C., north to Alaska, as well as the Queen Charlotte Islands and the east and west coasts of Vancouver Island.

With the development of new roads on the coast, and the advent of commercial aircraft with scheduled flights to most of the coastal communities, the need for so large a vessel diminished. In 1953, the *J. M. Swaine* was sold. Converted to a fishing boat by the new owners and later to a tug, she plied her trade in the coastal waters until 1976. Her demise was reported by the press as follows: "The three-man crew of the *J. M. Swaine* escaped unharmed when fire destroyed the vessel while towing a boom of logs from Powell River to Vancouver. Cause of the fire is unknown." So ended the one and only vessel owned by the Federal Department of Agriculture.

To replace the *J. M. Swaine*, we acquired a smaller, faster boat, a 34-foot gas-driven vessel named the *Forest Biologist*. It had the advantage of not requiring a qualified skipper, as it could be legally operated by regular staff personnel, and it served a valuable function until 1960.

There is much about the west coast that sets it apart from any other part of Canada, or, for that matter, any other part of the continent. The numerous inlets reaching, in some cases, 60 to 70 miles into the heart of the coastal mountains, the myriad islands, uninhabited beaches, coves, lagoons and secluded bays are generally considered to be intriguing, romantic, beautiful and intensely interesting. To me, however, the coast does not have all that much appeal. The climate is wet, and the water is generally very cold. There is not a great deal of difference between summer and winter. The hundreds of miles of uninhabited shoreline and islands testify to the generally unfavourable weather and difficulties of transportation other than by air or water.

Nevertheless, some people do not regard these factors as drawbacks, and enjoy the climate and the isolation. Two such people were Mr. and Mrs. James Stanton, of Knight Inlet.

We had travelled up Knight Inlet in the *J. M. Swaine*, and by late afternoon of a cold, rainy November day were anchored in a quiet protected bay for the night. There were five of us on the boat: Lew Fiddick, Howard Vey, Ken Hughes and the skipper, Bill Cleveland. Lew, Howard and I decided to take a walk up the beach for exercise, regardless of the rain. We had earlier spotted a log house back in the bush and decided to investigate, there being nothing better to do. As we came close to it, we could see it was occupied, so we knocked on the door in the hope the inhabitants might be glad to see a visitor on such a wet afternoon. The door was opened by a most gracious and hospitable woman who introduced herself as Mrs. Stanton. Nothing would do but we must come in and have a cup of tea. Seeing a big fire blazing in the fireplace on this miserable afternoon, we were delighted to accept her invitation.

Before long, Mr. Stanton arrived, and we soon realized our good fortune in meeting these two fascinating people. As an old-timer, Mr. Stanton was able to assist us greatly, both by providing us with information on the inlet and by volunteering to be our guide the next day. We wound up employing him at various times as a guide and bodyguard for the workers, since the area is noted for an abundance of grizzlies.

For two evenings, we sat before the Stanton's glowing fire and listened to their interesting stories. The log house had been built by the two of them, and contained the largest fireplace I have ever seen. The front opening was five feet high and eight feet wide, the fire bed five feet deep. It accommodated enormous logs and one could actually step inside when stoking it. It was nothing less than an indoor bonfire. They had, of course, an unlimited supply of wood, and once ablaze, the fire lasted all day.

The Stantons were from Seattle, Washington, where he had been in business as a car dealer. Growing weary of the business world, he and his wife had decided in 1919 to head for the wilderness. Together they headed north, up the coast of British Columbia, in a small boat. They were at that time in their early thirties.

Eventually they reached the northern portion of Vancouver Island and considered settling in the region near Salmon River. Fear of other settlers moving into the locality persuaded them to look for greater isolation. Knight Inlet was proposed to them as an area totally void of any habitation, with little likelihood of any changes in the foreseeable future.

They chose a site near the entrance of the inlet, but after a few years decided a better location would be at the extreme end, some 75 miles in from the coast. There they built the home in which we sat that evening. Mr. Stanton made his living by combining a little hand-logging, when he had an opportunity to sell his logs, with some commercial fishing. In later years, the reputation he had won as a guide led to his employment by wealthy tourists who came to Knight Inlet in their palatial yachts.

To Mrs. Stanton, grizzly bears were pets. She informed us she had three tame grizzlies which visited her daily, begging for food. Mr. Stanton had found them when only baby cubs, befriended and raised them. Now, as grown bears, they willingly came to her porch, where she stroked and fed them. One would occasionally walk right into the kitchen! Mrs. Stanton said she was extremely sorry they were not around so we could see just how friendly they were. With a very forced note of regret, we assured her we too, were sorry the three grizzlies were not in her house to greet us!

As we left to return to the boat, she reassured us of the gentle disposition of the bears. "If you encounter them on the trail," she said, "don't be alarmed. Just speak to them gently and they will be perfectly harmless."

This we agreed to do, but no sooner was her door closed than Lew exclaimed, "To hell with those three friendly grizzlies on that dark trail!'' We cut down to the open beach, the way we had come, and stumbled and splashed over rocks and boulders until we reached our boat.

The Stantons were interesting, both with regard to their past and their present mode of life. Their chief friends and neighbours were native Indians. As we sat before the fire in their home the second evening, a group of visitors arrived, eight Indians in all. Mrs. Stanton opened the door for them and they all filed in as if expected and sat in a row before the fire. Scarcely a word was spoken other than in terse reply to Mrs. Stanton's questions. She made them tea, which they all drank, this obviously being the usual custom. Abruptly, and without a word, they all stood and filed out of the door. Mrs. Stanton explained that they were her neighbours in Knight Inlet and were very fine and true friends. They would never pass her place without stopping in for a social visit.

Up to the time of our visit in November of 1947, the inlet had no settlers other than the Stantons. Recently, however, a forest company had set up a small camp at the head of the inlet and was doing some preliminary survey and cruising work. Although this camp was a long way from her residence, she remarked to us that scarcely a day passed when she didn't see a boat of one kind or other on the inlet. "A person simply doesn't have any privacy anymore," she sighed with despair.

Despite their isolation and loneliness, the Stantons were by no means unknown to the outside world. The climax of surprises came when they told us that during the latter days in the life of President Roosevelt, when he was suffering from the terrible strains of the war, his doctors had prescribed total rest and complete isolation from worldly problems and that as a suitable refuge, the Stanton's home in Knight Inlet had been selected. Just how this came about I never did find out, but it was probably through some wealthy American tourist whom Mr. Stanton had guided in the past. In any case, the Stanton's home was considered by the White House as an ideal retreat for the ailing President. An advance guard had arrived and all preliminary arrangements for the President's visit were made when he suddenly passed away.

With a three-whistle farewell to the Stantons, we left the head of Knight Inlet the following morning, November 13th, and headed for Chatham Channel en route south. Our skipper, Mac, a deep-sea man unfamiliar with navigating inside waters, accidently got us into Carrol Cove. Lew and I were in the wheelhouse at the time, looking at the chart. Lew whispered to me, "You know, I don't think we are in Chatham Channel. It looks more like Carrol Cove." I agreed that I, too, was beginning to question our course. Our skipper was a qualified man, in charge of a government vessel, and we hesitated to interfere with his responsibilities. About then, however, the skipper also began to doubt his course. Carrol Cove is shallow at best, but with an ebb tide, the possibility of being grounded had us worried. We decided to try a cautious turn about while we still had sufficient depth of water. Within a matter of seconds, however, we felt a soft but very firm drag on the keel and the ship came to a stop. We were aground.

Considering the rapidly out-flowing tide, we knew we had no time to lose. Ken, the engineer, and I launched our lifeboat and speeded to a nearby logging camp on Minstrel Island to ask for a tow from their tug in the hope they could pull us free. Because of the ebbing tide, however, they did not dare approach us for fear of being caught in the same trap. Their suggestion was that we should go ashore, cut poles, and prop the ship to prevent it from rolling over, as the inlet would soon be completely dry. Returning to the *Swaine*, Ken remained with the vessel while Lew and Howard went ashore to cut poles which I ferried across to the stricken vessel, where Ken and I installed them as props. Once we were confident the listing vessel would remain upright, all we could do was wait. The water around us continued to drain away as the tide ebbed, until at last there was no water in sight. We prepared supper as best we could with stove, table and floor all listing at a 30 degree angle.

About nine that evening, the tide began to flood, but with it came an onshore wind. The depth of water increased and the ship righted herself; the wind increased to gale proportions and blew us steadily further on to the sandbar. We were being forced dangerously close to some jagged rocks. We ran a line to shore from our stern to hold our drift to starboard, while Lew and I returned to the logging camp for help from their tug. They were now very willing to assist, as there was sufficient depth of water for safe navigation.

With a line from the tug, we let go our shore line. Instead of being pulled into safer waters, however, the wind now blew the *Swaine* further onto the sandbar and the tug dangerously close to those jagged rocks. The situation was worse than ever, and the tug was making no headway against the wind and the drag of the *Swaine* on the sand. Suddenly there was a momentary lull in the wind. It lasted no more than a few seconds, but it was enough. The *Swaine* was free, and both ships moved into deeper waters. Our boat could now move on her own, away from Carrol Cove and to safe anchorage for the night.

CHAPTER TWENTY-TWO

In 1950, the Kitimat smelter of the Aluminum Co. of Canada was still in the planning stage. It was proposed that the electricity needed for the smelter would be generated by harnessing the water power from a series of lakes in Tweedsmuir Park. These lakes lay on the east side of the Coast Mountains, but by damming the outflow of water on the Nechako River and constructing a ten-mile tunnel through the base of the mountains, the entire drainage system could be diverted to the coast to generate power at the proposed plant at Kemano. Electricity so produced would then be transmitted to the smelter to be built at Kitimat.

The lakes involved in this massive flooding proposal comprised the most important aspect of Tweedsmuir Park, those which permitted a circle tour through a wilderness waterway unique in its rugged beauty, its sport fishing and unspoiled solitude.

The principal dam, to be built on the Nechako River, would raise the water level at the dam site by 300 feet and flood the land for a distance of five miles on either side of the existing river at the dam. It would also raise the water level of the entire lake system, flooding the shoreline for varying distances according to the topography of the land.

There were, obviously, many potential side effects from such a drastic change in the Nechako drainage system, the most important of these being its effect on the downstream salmon population. This aspect was investigated jointly by the International Pacific Salmon Fisheries Commission, the Fisheries Research Board of Canada and the Department of Fisheries, Canada.

With respect to Tweedsmuir Park, its recreational potential, wildlife, sport fishing, timber and other ecological considerations, a separate investigation was initiated by the provincial government. This group consisted of: Mr. Russ Potter, Chairman, Fraser River Basin Board; Dr. Dave Turner, Deputy Minister, Lands; Dr. Pete Larkin, Limnologist, University of B.C.; Mr. Si

Oldham, Head, Provincial Parks; and myself as Head of the Federal Forest Biology Laboratory, Victoria.

An on-the-spot examination of the affected region was organized by a well-known guide and outfitter, Frank Hanson, who supplied boat, equipment, food and cook. Travel through the water course was by riverboat, a special adaptation for British Columbia lakes and rivers. Thirty-three feet long, three feet at the waterline, flaring to a six-foot beam at gunwale, flat-bottomed with a minimum draft and powered by outboard, the riverboat accommodated our party with ease.

For ten days we travelled through the lakes and their connecting rivers, which included Ootsa, Tahtsa, Whitesail, Eutsuk, Euchu, Chelaise, Tetachuk, Natulkz and Intata Lakes. Each night, after setting up our camp, we reviewed the day's observations and probable results of the proposed flooding.

On completion of our work, we submitted to the provincial government a joint report with recommendations. Nothing further was ever heard of it, nor was there any evidence that anyone ever read it. Certainly none of our recommendations was followed. It seemed evident our enquiry was nothing more than a process of going through the motions of an enquiry, the powers-that-be already knowing full well what was going to be approved.

Stipulations set forth by the International Pacific Salmon Fisheries Commission and federal government researchers recommended the release of water into the Nechako River in adequate quantities to give a sufficient flow and depth of water for fish transportation, spawning and rearing, and that a sufficient quantity of cooling water be discharged from the Nechako reservoir for fish survival and well-being (68 degrees F. in the Nechako River being considered as the high point beyond which the survival of sockeye salmon is severely impaired). None of these recommendations was included in the license issued Alcan by the provincial government. The license gave the Aluminum Company of Canada a "carte blanche" permit to harness the Upper Nechako River, with no consideration whatever to the salmon or any environmental effects.

In November 1979, water levels became critically low for salmon survival and the Federal Fisheries demanded the release of more water. Alcan contended that by their license from the government of British Columbia, only the Controller of Water Rights for the province had the authority to order such release and he had not done so. A court action followed. The decision of the court was that, by the Federal Fisheries Act, Alcan was required to supply "such quantity of water, at all times, as will, in the opinion of the Minister, be sufficient for the safety of the fish and for the flooding of their spawning grounds to such depths as will, in the opinion of the Minister, be necessary for the safety of the ova deposited therein."

The original permit of 1950 had been given at a time when there was little public awareness or participation in the matter of resource use or related environmental issues. Today, public opinion and interest is a critical consideration in any such negotiation. In this charged atmosphere of public concern, Alcan, in recent years, has become prominent in its recognition of resource inter-relationships and now places a high priority on the protection of salmon. That is evidenced by the most recent proposals made by Alcan to the Department of Fisheries and Oceans which include temperature control releases in the Nechako and Nanika Rivers, and compensation by enhancement for lost rearing capacity for chinook and coho salmon and rainbow and steelhead trout. All this would be aimed at a no net loss of fish production.

The harnessing of the water power of Tweedsmuir Park and the development of Kemano and Kitimat were economically and socially desirable, especially for the jobs the projects would provide. In spite of the recommendations from those of us entrusted with this responsibility, the provincial government totally disregarded the many assets of Tweedsmuir Park. Its recreational potential, its timber throughout the flooded areas, its integral part in the perpetuation of the salmon stock of the west coast, were all ignored. This treatment of an irreplaceable heritage was inexcusable and made a mockery of the oft-flaunted concept of multiple resource use.

Resources belong to the people. Developments such as the one in Tweedsmuir Park show that it is necessary for responsible citizens to act to protect our resources from being given away or destroyed for political or monetary gain.

Fortunately, some other resource users accept the general concept of multiple use as desirable and necessary. This acceptance applies particularly to some of the larger companies of the forest industry, who have given recreational and other related uses an important place in their long-term resource management plans. This is evidenced in their co-operative participation in the activities of fisheries and wildlife projects, and in their establishment of public campgrounds, boat ramps and road access. Nevertheless, there is still much to be desired from the standpoint of the industry's full acceptance and development of the multiple use concept.

CHAPTER TWENTY-THREE

Although there was no officially organized nor overall salvage of the doomed timber in Tweedsmuir Park, there was nevertheless one remote spot where a "gypo" logger known as "Blackie" worked with a small crew in an attempt to salvage what timber he could before being flooded out with the rising water. I visited his camp one day and stayed for the evening meal.

Blackie was an old-time logger, a big, burly, black-bearded, bull-of-the-woods type. And Blackie had a pet. He called her "Pansy." Pansy was a pig! How Blackie came to possess a pig even Blackie himself was none too sure. The story was that after a weekend drinking spree, Blackie had returned to camp with a pink baby pig in a sack. When he sobered up, he realized he had in his possession an animal, which when fed with waste from the cookhouse, would eventually supply his hungry crew of men with cheap pork.

Blackie's name, however, was no indication of his heart, for it took only a few days for Blackie and his pet pig to become inseparable. With the copious supply of food in the camp, Pansy soon blossomed into blushing girlhood. At the time of my visit to the camp, Pansy was six months old and I would estimate her weight at about 160 to 175 pounds! Blackie kept her scrubbed spotlessly clean. No pig could boast of a more thoughtful or attentive master.

In her baby days she had a pen, but on reaching her youth, she was allowed to roam free about the camp. On Blackie's return from the woods at the end of the day, Pansy would run to meet him, squealing all the way. He would greet her with a scratch behind her ears, and she would follow at his heels, gently grunting with each step.

At the evening meal during my very brief visit to his camp, I was seated with the crew at a long dining table with Blackie seated at the head. In the middle of supper, the white sow entered the cookhouse, grunting as she came, and proceeded to waddle under the table, between the feet of the diners, until she made her way to where Blackie sat. There she lay like a pet pup while Blackie tossed her bits of food to keep her happy.

Although the meal was excellent, and the camp clean, a pig in the cookhouse and the crew's complete lack of concern about it astonished me as I compared their tolerant attitude to what I would expect from loggers on the B.C. coast.

But Pansy's future was fraught with problems. Blackie's operation was due to wind up in the very near future as the waters continued to rise and he would have to leave.

Across the bay lived an old recluse. Pansy had been offered to him on condition she would be kept penned until after Blackie left. She was delivered to the old man but, exercising a pig's uncanny homing instinct, was back within a day, contentedly resting at Blackie's feet in the cookhouse. I asked Blackie just what he was going to do with her when he had to leave. "She's a real problem," he sighed. "If only Pansy was a dog."

*　　*　　*

In 1954, I was with four members of the B.C. Forest Service, one of whom, John Stokes, was a good friend of mine who later became Deputy Minister of Lands and Forests of British Columbia. Our mission was to survey the intensity of a very active bark beetle outbreak in the Flathead country of southeastern B.C.

We had laboured our way in two Land Rovers to an almost inaccessible region in the Flatheads of the Kootenays, and had selected a camping spot for the night. The weather had been threatening all afternoon. We had no tent, so three of our party prepared their sleeping bags in the open, wrapping them around with protective canvas.

John and I figured we had a better idea. We made our beds on the floor of an old abandoned log cabin which we had discovered a short distance away. Just as we all turned in, rain began to fall. John and I settled in our warm, dry sleeping bags with a certain smugness as we listened to the rain falling on the roof of the cabin, visualizing the others lying in the grass outside, rolled up in their ground sheets.

144

For a few minutes we lay in darkness and silence. Suddenly John was disturbed by something on his bed. "Hec," he called, "there's something running over my bed. I think we've got mice here."

"Yeah," I responded, "well, I suppose we should expect a few in an old shack like this. We probably won't notice them once we get asleep." In the ensuing silence, I recalled my previous experience with mice in the old cabin in the Rockies.

"Hec," John called again after a short interval, "there's not just one mouse, there's a whole herd running over my bed. We can't sleep like this."

"That's because they like you," I replied, turning on my flashlight. The mice disappeared.

I turned off the light and we dozed off only to be awakened by a loud gnawing sound coming from underneath the floor.

"That's a bloody pack rat," John sputtered as we both got up and pounded on the floor. This apparently frightened the critter away so we both returned to our beds.

After a period of peace and quiet, I was awakened again by John's hushed voice. "Hec, there's something else on my bed now."

"What, more mice?"

"Hell, no. This is big, feels about the size of an old tom-cat. Maybe it's a skunk! I don't dare move. Turn on your flashlight and tell me what it is."

I sat up in bed, and in the circle of light I saw the culprit, very much alive, big as an old tom-cat, sitting on top of John and staring me straight in the eye.

"John," I said, "I've got news for you. Your bed companion is not a skunk, nor an old tom-cat but a lovely big porcupine, and he's not the least bit frightened. Anyway, don't do anything to disturb him. I'll try to drive him away but you had better cover your head and face in case he takes off across you and your pillow."

The porcupine, body and tail bristling with long needle-sharp barbed-point quills, could drive hundreds of them into the sleeping bag or one's flesh with one slash of his tail. This was an even more formidable foe than a skunk. I turned off my light and for a moment we waited in silence, hoping for some action on the part of the visitor. Finally, in his characteristically slow and cumbersome manner, he moved across John's bed and onto the floor. With the aid of my flashlight, we now herded him out through the door, which we managed to close behind him in spite of a broken hinge. We then blocked the hole in the floor through which the porcupine had entered and returned to our beds.

No sooner were we settled than two pack rats took off across John's bed, then onto mine. With this John exploded, "To hell with any more of this," and jumping from his bed, sleeping bag under his arm, headed for the door. "I'll sit the rest of the night in the Land Rover, uncomfortable as it may be, but at least I'll be free of these bloody animals."

He disappeared through the door into the rain and darkness, while I spent the rest of the night in the home of the mice, the rats, and the porcupines, none of which I saw or heard again. They obviously preferred John!

CHAPTER TWENTY-FOUR

The year 1955 brought with it a marked change in my life pattern. I was transferred to La Belle Province, Québec.

The Federal Forest Research Laboratory was located in the School of Forestry at Laval University in Quebec City. For the next two years, my home was Quebec and my work extended through that province and to some extent into New Brunswick.

I found Quebec very beautiful. The autumn foliage of the sugar maple and other hardwoods is spectacular and for friendliness, hospitality and courtesy, the Quebecer is in a class by himself.

The city of Quebec is, in my opinion, the most fascinating in Canada, a city of endless discovery and adventure—places of historical significance, old dungeons deep within the city wall and other remnants of the past, many of which are unknown even to many of the local residents who have lived there all their lives.

Quebec City and Victoria have much in common. Both cherish their old traditions and landmarks, and rebel against the ripping down of old architecture for replacement with box-like glass and steel shells. Each is supreme in its own special category. Quebec is the most historic, Victoria is the most beautiful.

One very pronounced contrast between Quebec City and any other part of Canada with which I am familiar, is the general appreciation and support of the arts which I found evident in all age classes, occupations and professions. This was of particular interest to me as I had done a considerable amount of landscape painting over the years and had achieved some degree of success along this line. Three of my paintings had been hung in the Vancouver Art Gallery jury shows, two in similar competition in the Victoria Art Gallery and one is permanently hung in the Vernon City Hall. While in Quebec, I enrolled in winter courses at L'Ecole des Beaux Arts, which is affiliated with the University of Montreal.

At this outstanding school, I registered for the night courses comprising three classes per week over six months, at no charge; I had a one-man show of 45 paintings in the civic auditorium Le Palais Montcalm, 2:00 p.m. to 10:00 p.m., for a week, for which there was no charge; sale of paintings, handled through the front office, also involved no charge or commission. Art classes were filled to capacity with the students fairly evenly divided between men and women and practically everyone in their twenties. In my experience elsewhere, the opposite usually applied on every count.

My official work was largely administrative and while it gave me an excellent opportunity to see the province, it did not really appeal to me. The main research work related to the notorious spruce budworm of eastern Canada. Aerial spraying had been in progress in New Brunswick for several years, but Quebec had withheld such action, despite heavy losses to their forests. It was finally decided that applied control was essential, and to this end, a field research station was set up on the Gaspé Peninsula at New Richmond, an area which had been predominantly English in the early years. This was magnificent farm land which had recently become a liability, incapable of producing a living for the owners because the traditional market for farm produce from the Gaspé had been the eastern American cities, and the trucking of American grown farm produce had totally eliminated these markets for Quebec growers. Almost all the farms had been abandoned.

Despite the fact I had taken French for three years in my high school days, had passed my Ph.D. French requirements (translation only) at McGill, and was associated with French-speaking people at the research laboratory at Laval University, I never became sufficiently proficient to carry on any kind of a conversation with a French-Canadian. This was due in part to the fact that whenever I tried to make myself understood in French, my conversational partner would immediately start speaking English, since it was easier for him than trying to understand my fractured French! My struggle to master the French language came to a climax one day when I gave a ride to a Catholic priest I found trudging down a dusty road on a hot summer afternoon. In my very best French, I asked him where he was going. He attempted to answer me, so I thought, but I was unable to make out anything he was saying. Over and over I tried to get better directions and again and again he tried to explain himself. The only thing clear about this conversation was that neither of us knew what the other was saying. In desperation, he finally asked, "Vous parlez anglais?"

"Yes," I said, "I certainly do."

"Well, so do I and I don't speak French! I'm visiting here from Regina."

*   *   *

It was during our second year in Quebec that two events of major personal importance occurred. The first concerned our daughter Donnie, the second Vi and myself. The latter was to open an entirely new era in our lives.

During our residence in Victoria, Donnie had been an ardent skater who worked long and hard to perfect her art. Her ultimate ambition was to join the show, "Ice Capades." Skating every morning at six, again each afternoon, winter and summer, she brought her skill to a truly professional level.

With the appearance of "Ice Capades" in Victoria, auditions for hopeful stars were conducted. Donnie was auditioned but we heard nothing further. The event was more or less forgotten. Then came our move to Quebec.

That autumn, Donnie entered Macdonald College of McGill University. No sooner was she settled in than a telegram for her arrived from Phoenix, Arizona. We read it with curiosity and then building excitement. Ice Capades wanted her to go immediately to Phoenix and join the company. We phoned to relay the exciting news.

"But I can't go now," she replied. "I've already paid my fees."

For this young girl, who had dreamed of such an opportunity for so long to be suddenly confronted with the real possibility was indeed a shock. Hoping and striving for something that seems to be almost unattainable is quite different from having to make a firm, rational decision about a real situation.

"Forget the fees," I told her. "This is your life and the decision must be your own. Take your time, but remember that what you decide will probably affect your entire future. When you decide,

149

discuss it with your mother and me before answering this telegram."

Donnie gave the matter careful thought, and in the end rejected the offer. This, however, did not terminate the question. Ice Capades countered with the suggestion that she should go to Hollywood the following June to join the company as they prepared for a new production. This time, Fate stepped in and settled it once and for all.

News of the Ice Capades offer soon spread throughout the McGill student body. When an ice carnival was organized by the university, Donnie was selected as one of their principal skaters. During the performance, while in the midst of a spin, her blade caught in a crack in the ice. With a twist of her knee, she collapsed.

Although the university sent her to the best specialists in Montreal, none could assure her that an operation would completely repair the knee. There was, in fact, danger that an operation could result in a stiff knee joint. On the other hand there was the possibility the knee might repair itself, if left alone. The latter course was followed. In due time, the knee recovered but the accident put an end to further serious figure skating.

The second event was to change completely the future of all of us. We had just completed twenty months in Quebec when a letter came to me from MacMillan Bloedel Ltd., in Vancouver. The letter asked if I would be interested in leaving the government service to return to British Columbia and accept a two-year contract with the company to work on a very serious insect problem affecting their high value saw and peeler logs. The offer went on to suggest that on completion of the two years, I could remain with them on a consulting basis but would also be free to undertake consulting work with any other interested firm.

The idea of returning to the Pacific Coast and the province of British Columbia appealed to both Vi and me. For several days we gave the proposal much serious thought.

After more than 30 years service with the federal government in the field of forest entomology, I had acquired a great many friends and associates in both Canada and the U.S. To cut these ties and move into an entirely new field of work with the industry was hard to contemplate. Too, I would be leaving the government service at a time when pension benefits were at their greatest rate of increase. Eventually, I made up my mind and with mixed emotions, wrote to my superior in Ottawa, Dr. M. L. Prebble, to inform him I was resigning from my job.

Then, of course, I began to worry. Had I made the right decision? Today, more than 20 years later, I look back and realize the move was very fortunate for me, but I still wonder, with some degree of amazement, how I had the courage to make such a major move. Certainly without Vi's support, I could never have done it. It is with a great sense of gratitude and good fortune I recognize how so much has come my way without any special plan or design.

There was a two-month period between my departure from the federal service and the beginning of my work with MacMillan Bloedel in B.C. We took advantage of this interlude by purchasing a trailer in which we travelled down the eastern seaboard to Florida, around the Gulf of Mexico to California, and up the west coast to B.C.

*   *   *

Reporting on the first day of January 1957 at the Vancouver office of Canada's largest forestry company, I was directed to Nanaimo on Vancouver Island, where I would be affiliated with their Forestry Division. Thus, it came about that I purchased my present 35-acre waterfront property while land prices were still at a reasonable level. At the forestry headquarters in Nanaimo, Ian Mahood, Forest Manager, told me what would be expected of me.

"There's your desk. We don't care how much you use it, nor how often we see you. All we want are definite and final results at the end of your two-year period."

With instructions like that, I could not have had a freer hand to organize a research programme, nor a more uncompromising demand for positive results at the end of the project! Although the programme reached a successful conclusion, it was not without its share of the unpredictable biological problems bound to arise in the course of such work.

Part of my job was the exploration of various ways and means to reduce or control the damage caused by a small wood-boring insect, the ambrosia beetle, which attacks and bores into recently felled logs. My work also included the determination of dollar losses to lumber and plywood through studies in the mills; the reduction in damage to logs in the woods through modification in woods management involving all the company's operations; the beetle hazard to logs in storage both in water and on land, and the experimental control by insecticides applied by aircraft to stored logs. Some years later, when I was working under the sponsorship of the Council of Forest Industries of British Columbia, a continuation of this work led to a system of water misting over stored logs which gave total protection without using any chemicals or insecticides. Subsequent research undertaken jointly between scientists at Simon Fraser and U.B.C., has opened an entirely new field of control through the use of pheromones (scent produced by the animal), as a non-chemical means of control. (See Chapter 29)

The ambrosia beetle attacks and bores into green logs soon after they have been felled. Small tunnels are projected and a black fungus, on which the insect feeds, develops within the tunnel. The spores for this fungus are carried on the body of the female. The beetle literally carries its source of food for itself and its young wherever it goes. It does not feed on the wood. The drilled tunnel is merely a home in which to live, the excavated borings being pushed out of the entrance hole, becoming very noticeable on the bark of the log. Eggs are laid within the tunnel. The young, on reaching maturity, fly away in search of a suitable log in which to bore, and another generation of ambrosia beetles comes into being.

The damage which results from these borings shows in lumber and veneer as blackened holes, tunnels, niches and other defects. There is a loss of hundreds of thousands of dollars annually of otherwise high quality lumber and other manufactured products. The greatest impact of the ambrosia beetle is found on the export market where there is strong resistance and in some instances an outright embargo against importation of forest products showing any degree of attack by this insect.

The adult beetle flies to and infests green logs both on the land and in water storage. Although logs can be protected from attack by the use of chemical insecticides, the procedure is not entirely practical because the insect attacks on all sides of the log and it is physically and economically impossible to cover the undersurfaces of a log when it is lying in the woods. The use of chemicals was therefore restricted to logs floating in the water. In such an instance, only the upper surface is exposed to attack and that can be readily sprayed by helicopter. The programme was developed in co-operation with the Federal Department of Fisheries and in time became an annual project and the only practical method of coping with the problem.

The acceptable insecticide in use against the ambrosia beetle in other countries was Benzenehexachloride, commonly referred to as BHC. BHC, however, is a close relative to DDT with many of the same undesirable characteristics which made it an outlawed insecticide over much of the world. Hence, it became imperative to find an alternative to BHC which would give the desired protection, yet be environmentally compatible.

After experimenting with many potential substitutes, a suitable alternative (Methyl trithion) was chosen. It fulfilled the specifications, was totally harmless to fish in the dosage and manner used, so far as our work could establish, and was as effective as BHC against the insect. It did require extremely careful scheduling in use, however, since its life was relatively short. Since this chemical could achieve the desired results if accurately applied, further use of BHC was discontinued.

In spite of improvements, there evolved a growing opposition to the use of any insecticides over fresh water lakes, and in 1970 the Provincial Department of Health ruled there should be no more application of poisonous materials on any fresh-water lake in the province. They had no objection to further use of Methyl trithion on salt water though, due to its rapid breakdown and wide dilution in the sea. This ended the spraying of logs in fresh water. By this time the IWA had become involved, and they refused to handle any logs sprayed with insecticide either in salt or fresh water. Faced with mounting opposition to the spraying of logs, the forest companies decided to abandon further use of chemicals and instead concentrated on non-chemical measures for prevention of ambrosia beetle attack.

Studies by the Federal Forest Research Institute in Victoria indicated that the time of year when trees were felled had a great influence on the degree of susceptibility to future attack. It was found that a period of "seasoning" was needed after the tree had been cut before it became attractive to the insect. This "seasoning" period is usually a few months, depending on the season of the year and the prevailing temperatures. Fall- and winter-cut logs were always the most severely infested with the commencement of insect activity in the spring.

A practical method of woods management was developed whereby logs were removed from the woods and processed on a carefully prepared schedule before they reached the susceptible stage for ambrosia beetle attack. Important strides have been made in many operations in the carefully planned utilization of logs as a means of reducing beetle damage, but in recent years, there has evolved a new technique of handling logs which further complicates the problem. This is known as "dry-land sorting."

As the name implies, the logs are no longer dumped into the salt chuck but instead are held on dry land, where they are sorted, graded and stored in decks 15 to 20 feet high over many acres of land. This creates a situation whereby the insect has ready access to the entire inventory of logs. Protection by chemical sprays is physically impossible, since the beetle attacks throughout the pile. With the advent of dry-land sorting a new and different approach to log protection became imperative.

Previous studies had shown the ambrosia beetle to be readily influenced by moisture and temperatures of the air. Changing the atmospheric conditions, therefore, offered possibilities for repelling the insect from the area. I worked in co-operation with an irrigation specialist, W. Calder, and a hydraulic engineer, M. Polham, to produce specifications for a technically sound installation. The objective was to saturate the air in the storage area with a water mist or fog, using a minimum of water under high pressure, making the region undesirable for the insect. If effective, the system would comprise a permanent underground installation with 15-foot risers and nozzles capable of being removed from the underground water mains as desired to allow normal operations within the area without disruption of the fogging equipment.

After one season's trial with a pilot model, a detailed biological assessment was undertaken by the Federal Forest Research Centre. Results proved the system to be 100 percent effective, using only sea water.[2]

The installation of a full-time operating unit in the Renfrew Division of B.C. Forest Products has resulted in the complete control of the ambrosia beetle in their dry-land sorting grounds.

While the story of the ambrosia beetle ends on a happy note of partial solution, having met the combined opposition of government, labour and the public, it is, nevertheless, far from finished. This persistant situation, like all native insect problems, is with us for all time and is something we have to live with.

Much of the credit for the success of this long and costly project belongs to the support given by two outstanding foresters: Grant Ainscough, Vice-President, MacMillan Bloedel Ltd., and Gerry Burch, Vice-President, B.C. Forest Products. The experience also points up the value of opposition as an essential ingredient to progress, even though alternative solutions are slow, costly and at times frustrating.

*   *   *

While in the woods on the ambrosia beetle project, I had a memorable encounter with a cougar. The big cat, like the wolf, is generally regarded as harmless to humans, although there are authentic records of people being attacked by starving cougars. A cougar seldom exposes himself to view and this was the first and only occasion in which I saw one in the wilds, although I have no doubt that many have seen me.

I was walking along an old long-abandoned logging road. Ahead of me, an adult cougar stepped from the brush to the road. Surprised at my presence, she stopped for a moment, looked, then crossed the road and disappeared in the bush on the opposite side. Immediately following her, two cubs appeared. Equally surprised at seeing me, they paused, stared and then sat down in the middle of the road to keep an eye on me. Mother cougar must have told

[2] Richmond, H. A. and Nijholt, W. W., 1972. "Water misting for log protection from ambrosia beetles." Canadian Forestry Service, Information Canada. Pacific Forest Research Centre, Victoria, B.C.

them that staring at a stranger was bad manners, for suddenly
they both jumped up and disappeared into the bush to join their
mother. No sooner had they vanished, than a third one appeared.
He went through the same procedure, stared, sat down until
called, and immediately took off to join the others. Triplets, I
thought, and was wondering how I might get a better view of them
when two more appeared. The performance of the others was
repeated by these two before the mother and her family of five
vanished from sight.

Upon checking later with the game warden, I was advised that
five cubs was not an unheard-of number, although I was the first
person he had ever met who had actually seen that many.

CHAPTER TWENTY-FIVE

After completing my two years with MacMillan Bloedel Ltd., I was retained by the B.C. Loggers Association, which later became a part of the Council of Forest Industries of British Columbia. The Council is comprised of most of the companies within the forest industry of the province. Although I had a desk and phone in the COFI office in Vancouver, I was seldom in the city for any length of time, since my responsibilities were divided among the operations of many companies. My association with the Council was most satisfactory and enjoyable since the people I was associated with were among the finest with whom one could hope to work, and I was given a free hand to analyze current issues affecting the industry which in my opinion needed direct action or initiative, support or co-operation with other research agencies.

Up to this point, my life had been devoted largely to insect research and surveys with the federal government. My work now related directly to the interests of the forest industry. The principal concerns of industry are naturally protection of the crop and solving the question of how best to reduce damage caused by insects through practical and environmentally acceptable means. I enjoyed a very close liaison with the Pacific Forest Research Centre of the Canadian Forestry Service, of which I had been a part for so many years, with the B.C. Forest Service; the Fish and Game Branch of the provincial government, and with the Federal Department of Fisheries. These contacts were of inestimable value to me in my work with the forest industry, since few projects, if any, do not inter-relate with these agencies in one way or another.

*   *   *

In 1964, an insect infestation of the green striped looper developed on Graham Island of the Queen Charlotte Group. The infestation, although not widespread, had the potential of becoming a serious problem. We recognized this as an opportunity to try some experimental control with newly-developed insecticides which might be more compatible with the environment than materials used previously. I was, therefore, on the Islands for the better part of one summer, during which time I had the opportunity to become acquainted with that unique part of Canada and with some of the people residing there.

The Queen Charlotte Islands are located some 50 miles west of the nearest mainland point of northwestern British Columbia. The weather pattern in the region contributes to their isolation and lack of appeal to many as a place for permanent residence. In the early part of this century, however, thousands of acres of "promising" farm land were sold sight-unseen to adventurous Englishmen and other settlers. The financial panic accompanying the depression of 1907, and the lack of jobs in the cities had created a great upsurge in the movement of the jobless to the land.[3]

The British Columbia government advertised the wonderful farm land available on the Queen Charlotte Islands through widely-distributed brochures. These Islands, they said, were destined to become a bread basket and supplier of farm products to the metropolis envisaged for Canada's western gateway to the Pacific, Prince Rupert. If the applicant was a British subject or had made application for naturalization, he could pre-empt 160 acres. The only other requirement was that he live on the land continuously for six months out of each two years, do $400.00 worth of improvement work and pay $1.00 per acre and a Crown Grant fee of $10.00. The land, however, was not surveyed or mapped, a fact which ultimately led to a great deal of confusion.

An aura of optimism about the future of the Charlottes was evident as early as 1904. In March of that year, the *Victoria Colonist* reported: "The north end of Graham Island is making great progress and settlers are pouring in with each boat." One report stated that the soil was so fertile that two sacks of seed potatoes would produce 2,600 pounds of new potatoes. A news-

---

[3] Dalzell, Kathleen, 1968. *Queen Charlotte Islands, 1774-1966.* C. M. Adam, Publisher, Terrace, B.C.

paper report in 1905 stated "Graham Island has the finest soil in British Columbia, with room for hundreds of families. Timothy hay grows six feet tall and there are thousands of acres with not a bush or tree upon it—deep, rich, black peat and sandy loam."

A particularly optimistic and imaginative venture was reported in the *Victoria Times*, May 1907:

"Three thousand acres on Masset Inlet have been purchased by a Winnipeg company on which is proposed a townsite to be called Graham City. Consequent upon the enormous development work well on the way on the Queen Charlotte Group, many towns will spring up through the district, not the least among these will be Graham City, the future headquarters of the Graham Island Steamboat, Coal and Timber Company. When their two large sawmills and coal mines are in full operation, they will employ three thousand men the year round.... Once Prince Rupert is established, Graham City will become a fashionable summer resort of the north. ... A Los Angeles capitalist will shortly establish a line of steamships to run between Graham City, Prince Rupert and points of importance on the Queen Charlottes and the mainland."

Further evidence of the optimism which prevailed is indicated by the advertisement appearing in the following year:

"THE REAL LAST WEST. The time has come when the real Last West has been reached. Behind are the valleys into which settlers are pouring at such a rate that it is only a matter of a short time when the ground floor of opportunities will be so well occupied that those who come later will have to take second choice. The heart of the Last West is to be found in Queen Charlotte City. Best investment in the province. Address: Queen Charlotte Townsite Co., Queen Charlotte City."

The Prince Rupert newspaper, *Evening Empire*, on September 11, 1909 reported that the town of Graham City had been re-named Masset. Although Graham City never developed as envisioned by the promoters, many settlers did come; farm land was procured and cleared and homes were built. Doubtless many years of backbreaking work went into the cultivation of these farms. Then came the First World War. Between 1914 and 1918, the population slowly diminished and one by one the farms were abandoned. Few, if any, returned and the great land boom died with the war. The great metropolis of Prince Rupert never materialized, and the weather of the Charlottes further discouraged attempts at permanent settlement.

While mapping by helicopter the extent of the insect infestation in the timber of Graham Island, I had an exceptional opportunity to see many of the derelict homes left over from the settlement boom. Their collapsed roofs, rotting floors, fallen fences and acres of what was once cleared land remain as testimony to the ambitions and dreams of those early pioneers. Many of the homes are accessible only by helicopter since any roads which may have existed have long since disappeared, taken over by forest growth.

In contrast to these crumbling remains, hedges of rhododendron, azaleas, lilac and other shrubs have flourished despite years of neglect, some attaining heights of 30 to 40 feet. I was fortunate to see them at the zenith of their summer blossom when they brightened and enlivened an otherwise melancholy cenotaph of bygone years.

Despite the comings and goings of man, one permanent resident has remained through the eons, the bald-headed eagle. Looking down from our helicopter, we observed one of these huge birds perched on the side of her nest. Within the nest were two large white eggs. We paused in our flight and, bit by bit, lowered the aircraft to get a better view. The nest was built on top of an old, windblown, weather-beaten, twisted tree leaning outward precariously over the edge of a sheer cliff, a drop of a thousand feet or more below it. The eagle watched motionless as the strange contraption hovered above her. She showed no fear; she seemed to know her solitary habitation was impregnable. From this lofty site, she reigned majestic over her wild domain, millions of acres of timber which greened the mountainous wilderness below.

Peculiar to the Queen Charlotte Islands are the wild cattle which range over eastern Graham Island. Early reports on file in the Provincial Archives state these cattle were first introduced on the Queen Charlottes by the Hudson's Bay Company prior to 1850. They were mostly Shorthorns but became crossed with Angus and Jersey from cattle brought in by settlers. The report estimates the population at 1,000 head. In their wild state, they developed long legs and lithe bodies and became fleet as deer. They have persisted through the years, their numbers doubtless controlled to some extent by weather, feed and in-breeding.

Although they are rarely seen, it was my good fortune to come upon them by helicopter in the course of insect infestation mapping. There were 40 or 50 of them and they appeared to be thriving, with some large and powerful animals amongst them. When we dropped down for a closer look, they panicked, colliding and leaping over one another in their frenzy to seek safety in the protective woods.

The *Vancouver Province* of 1904 tells of a Mr. James Prescott of Seattle who attempted to rope and capture some of these wild creatures, but whose efforts ended in failure. Undaunted, he planned to return the next year with some riders from Washington. He proposed to construct a mammoth corral and round up the entire herd, estimating the cost to be about $4,000.00. The plan seems to have collapsed as nothing further was ever printed about the venture.

The cattle remained in their wild state and in 1907 the settlers in the vicinity of Masset requested the provincial government to undertake a programme of eradication. The *Vancouver World* reported, "The cattle run wild, destroying crops and have viciously attacked both men and women on sight. They are, indeed, declared to be more dangerous than any other wild animal in British Columbia."

Through the intervening years, nothing has been done toward their elimination and today, wild cattle still roam at large on Graham Island.

*   *   *

During my stay on the Charlottes, I became acquainted, quite by chance, with one of the surviving families of the early settlers. I was working on a joint research project with the Industry and the federal government, with a long-time friend of mine, Jim Kinghorn. Jim and I had set out by boat to examine some sample plots situated on the north side of Masset Inlet. Masset, unlike most other inlets, leads in from the Pacific by a long, narrow neck of water, some 25 miles in length. This river-like neck expands into a body of water about 20 miles long and 7 miles wide. During the change of tide, the water rushes through the narrow neck like a swift and powerful river.

Jim and I were using a 15-foot aluminum boat with flotation tanks, powered by a 40-horsepower motor and were approaching the north shore when a heavy wind developed. We decided to make shore, hoping the storm would eventually abate. As we approached land, Jim spotted some log structures in the woods which we thought might be another old, deserted settlement worth looking over.

We were no sooner ashore than to our surprise, three young people appeared from the woods, two girls and a boy. Unlike most children who have grown up in isolation and who usually run and hide at the approach of a stranger, these children were very friendly and extremely polite. They were most anxious that we should go to their home with them where their mother would be glad to make us a cup of tea. They told us their names and ages, the boy being 14, the girls 7 and 11. We accompanied them to their house where we met their mother and father, Mr. and Mrs. Ainsworth. They had a small farm and he also did some salvaging of drift saw logs for a logging operation on Masset Inlet. In the course of the afternoon, we learned the mother's parents had come from Germany at the turn of the century, settled on this spot and proceeded to develop a farm. The mother was born there and apart from one occasion when she had been to Prince Rupert, had never been off the island. She spoke with a very pronounced German accent since, of course, she had no one but her German parents to teach her English until she married. Her parents had died some years ago.

The three children were taking their schooling through a Provincial Department of Education correspondence course, and were doing very well. The only association they had with other children was once a year, when they were taken to a Christmas party at Port Clements, a small community on the south side of Masset Inlet. The children were eager to talk, and gave Jim and me an escorted tour over their farm.

As the day wore on, the storm showed no sign of slackening. Mr. Ainsworth thought it might subside by sunset, as was frequently the case, so we were invited to stay for supper, an invitation we were very glad to accept.

When we sat down for supper, we noticed Mrs. Ainsworth and

the girls had changed to dresses, while the boy and his father were wearing clean white shirts. The table was spread with a linen cloth and napkins. Jim and I, in our dirty old bush clothes, felt rather out of place.

Supper finished, Jim and I insisted we attempt to get back to our camp, although the storm hadn't blown over. No one at the logging camp where we were stationed knew exactly where we had gone, other than that we were on Masset Inlet. Our failure to return would most certainly prompt a search by helicopter the next morning and cause considerable alarm all around. While Mr. Ainsworth thought a crossing might be possible, he strongly urged us to remain with them for the night. We were determined to attempt the journey, though, and so, with the motor refueled, we buckled on our life jackets and bade our new friends good-bye. With lukewarm enthusiasm, we headed into the storm.

All went moderately well for the first part of the journey, since our course took us on the leeward side of a couple of islands which protected us and gave us a false sense of confidence. No sooner were we beyond the shelter of these islands than we found ourselves in the centre of the storm. In addition, there was a strong tidal flow, which at this time of day, was at its peak. The wind was roaring at us from the opposite direction to the flow of the tide. These two opposing forces whipped the sea into a frothing frenzy in which we were totally helpless. One moment, the stern was practically buried with water coming within a fraction of the transom while the bow pointed skyward; the next moment, we were plunging headlong into the trough, the bow scarcely able to plough its way through the oncoming waves and water sweeping to the level of the gunwale. To put it very mildly, we were scared. It was already growing dark and it was obvious the storm would continue through the night. Jim, seated in the stern and handling the motor, shouted above the noise of the wind and the water, "What the hell are we going to do? Should we attempt to turn back?"

"I don't know, Jim! I've never been in such a fix as this. I think we'd best just keep our bow into the wind and hope we ride it out."

"Okay, but if she starts to take on any water, you bail while I try to hold her straight."

Our greatest fear was being swamped and capsized. We decided that if this happened we would tie ourselves to the boat and let the motor go to the bottom, hoping we might eventually be found. We realized, however, it would be impossible to survive in that cold water very long. Those thoughts in mind, we somehow held our own against wind and sea, being blown, very slowly, toward the opposite side of the inlet. This was certainly not by design nor through any manoeuvring on our part, nor were we even aware of it for a long time. Finally, however, we began to realize the heavy seas were being left behind and the tide was not as forceful as it had been. We were being blown out of the storm and across to the other shore.

"Now," shouted Jim, "Let's try a turn, run with the wind and head home." With the land engulfed in darkness, we reached our destination, very tired, very cold and very thankful!

When I later recounted our experience to the operator of the ferry between Graham and Moresby Island, he credited our survival to the fact that we had been too scared to attempt to do much of anything. Most tragedies, he believed, resulted from people who panicked and tried to force their boat against the sea, resulting in it being swamped and capsizing. But when I told Mr. Ainsworth the story of our perilous crossing, his reply was simply, "I guess your guardian angel was very close to you that night."

That was not the only memorable boating experience I had while on the Queen Charlotte Islands. The Islands are separated from the mainland by Hecate Strait, 80 or 90 miles of the roughest, windiest and most difficult stretch of water on the B.C. coast. To anyone familiar with this strait, the idea of crossing it in a bathtub is not only ridiculous but utterly impossible. But I did, or part of it at least. This remarkable feat occurred some years earlier while I was still with the federal government.

We were en route from Prince Rupert to Sandspit across Hecate Strait in our 60-foot vessel, the *J. M. Swaine*. It was an extremely rough and stormy day. Although she was quite capable of weathering the storm, the ship was taking a battering. From the crest of each mountainous wave, she would plunge nose-first into the trough to be hit by the oncoming wall of water which would batter the wheelhouse and sweep the deck as another wave approached. This continued hour after hour and we did little more than hold our own.

Unfortunately for me, my stomach was not made to weather such unnatural motion, and I became desperately ill. I hoped to find some degree of relief by rolling up on my bunk and keeping warm. No sooner was I down than it became necessary for me to make a dash for the deck! To get there, I had to pass through the engine room, where the smell of diesel only added to my misery. From there, it was up the ladder, through the hatch and across the deck, all of which took precious time, and time was what I did not have! After a second unsuccessful attempt, I decided my best course would be to stay on deck, but here I ran into a new problem. With each wave, a foot or more of water swept across the deck. I was getting tired of jumping up on something to escape the flood every time we hit a wave, so when I spotted a galvanized washtub hanging on the wall, I took it down, set it on the deck and cramped myself into it in a sitting position. The water now would sweep the deck, coming almost to the tub's rim but never running into it. My stomach continued to revolt against the situation, but at least I no longer had to worry about making it to the deck in time. Thus I travelled the rest of the way across Hecate Strait in a bathtub, establishing, I suppose, some sort of a record for bathtub travel! By the time we reached the calm waters of Skidegate Inlet and Sandspit, I was so cramped I could scarcely straighten my legs, so cold that I could barely feel them, and so sick I didn't care. And that's how I came to make this historic journey in a bathtub.

CHAPTER TWENTY-SIX

In 1960, an infestation of the blackheaded budworm (*Acleris gloverana*) on the Queen Charlottes led to another joint undertaking between the Council of Forest Industries of B.C., the Federal Department of Fisheries, the B.C. Game Branch and the Canadian Forestry Service. Although there was no official ban against the use of DDT at that time, we already considered DDT totally unacceptable for use as an insecticide against forest insects, so we were experimenting to find a substitute. This project was centred on Moresby Island, one of the largest islands of the Charlottes.

It was necessary that I rent a car for use while carrying out my work on Moresby Island. This transaction, together with a few others, revealed to me the very unusual business attitudes typical of the Queen Charlottes. Everyone seemed to trust everyone else, without question.

Separating Moresby and Graham Island is Skidegate Inlet. The only car rental agency was a service station on Graham Island. The car was eventually sent over to me via an old rusty wartime landing barge, which operated as a ferry between the two islands. Upon arrival of the car, I phoned the renting garage for instructions regarding payment of rent, gas, etc.

"It's all quite easy," the owner informed me. "Look in the glove compartment of the car and you will find a key and a book. When you need gas, go to Barney's gas pump, the only one on Moresby Island, unlock the pump with the key, take as much gas as you need and enter the amount in the little book. At the end of the month, tear the page out and mail it to me. Then I'll mail Barney a cheque for the gas you have used and I'll add that to your car rental for the month."

"But how does Barney know I have entered the correct figure in the book?" I asked. "He also has no check on who else might be taking gas with the use of the key."

"Oh," he replied, "We've worked like this for years. We never have any trouble. No one would cheat like that."

On another occasion, I was trying to call Vancouver from Moresby Island on a pay phone. The operator, speaking from the telephone office in Charlotte City on Graham Island, interrupted my call to tell me that the pay phone I was using was out of order and offered to put the call through for me. "How will I pay for it?" I queried.

"Well, I'm not exactly sure, but anyway, who are you?"

I explained who I was, and what I was doing. After a few moments of thought, she said, "You sound all right, so I'll tell you how we can work this. Do you know the fellow who runs the road grader on Moresby?"

"Yes, I know him to see him. I saw him working down the road this morning."

"That's good enough," she replied. "I'll put your call through to Vancouver and when you are finished, call me back and I'll tell you how much you owe. Then you find the road grader, explain it to him and give him the money. He'll give it to me the next time I see him." And so my call was put through with a degree of efficiency apparently normal by Island standards.

While working on the experimental spray project, I had a crew of four men for flying and aircraft maintenance. Because their working hours were so irregular, I had arranged that they could eat anytime they wished at a small restaurant near the landing field. It was agreed the proprietress, a lone woman, would keep track of all the food they ate and submit the account to me.

Weeks passed, but she never billed me. Finally, our work was completed, and in preparation for leaving, I asked for her bill. She presented me with a cardboard box full of slips on which was written the various orders. These were on scraps of paper, pieces of cigarette wrapping, napkins, bits of cardboard and the like, but nothing like a proper order form. She had obviously never heard of such a thing. "You can add those up and that's the amount you owe," she announced.

Accordingly, I took the mess, arranged the scraps of paper, listed and numbered each for cross reference, and when I had finished, presented her with a final statement and supporting entries. "What's all that about?" she asked.

"Well, I thought you'd like to check the statement and my figures."

"What on earth for? If you can't add that up right, I'm sure I can't. Anyway, you wouldn't cheat me. There'd be no reason why you should."

During my stay, a lot of drilling for oil was in progress in the Charlottes. Results were never made public, and eventually the drilling ceased. When I discussed the potential for oil development with one of the businessmen in Charlotte City, he expressed the feelings of many others in the community.

"I hope they never find oil on these islands. We are happy and contented as we are. We have nothing to gain by such development. Bring in outside competition from the cities and they would run us out of business overnight."

From my experience on the Queen Charlotte Islands, I am sure he was right. I have never known business to be run on a more delightfully trusting basis than it was when I was there. I could only wonder how long such a way of life could persist.

CHAPTER TWENTY-SEVEN

My more than 30 years experience with the federal government and eighteen years with the forest industry gave me an opportunity to compare these two fields of employment. Contrary to uninformed public opinion, it was my experience that the administrative efficiency of government equals that of industry, and in some instances, surpasses it. There can be as much red tape in industry as in government. Red tape seems to be an unavoidable consequence of bigness. A publicly-owned company, operating under competent management, can be operated as effectively as it would be in the private sector. The unfortunate impression the public has of government, engendered by political bureaucrats and unqualified political appointees, is in no way indicative of the dedication, sincerity and loyalty of the average public servant. With reference to professionals employed in the public service, the only group of whom I can speak with experience, I encountered a high degree of personal involvement. I believe this is due to the fact professional employees in government service plan to make it their lifelong career, with little thought of changing. On the other hand, professionals in private industry are inclined to be much more transient and frequently move from one organization to another whenever opportunity for advancement presents itself.

It seems to be the general belief of the public that decisions made by industry are immediate, definite and arrived at through the most efficient and direct means. In many instances, this is true. Conversely, citizens feel government service decisions are indefinite, slow and arrived at by round-about, ineffectual means, taking their toll in time, money and frustration. In many cases, this also is true. However, the private sector can be worse than anything I have ever known in government administration. In other words, in my experience, there is no "golden boy" in administrative efficiency in either government or private industry.

*　　　*　　　*

I had been closely associated with the control of the spruce budworm while in the East, though control had never been my principal interest in entomology. To me, living insects are much more interesting than dead ones. If, however, one becomes involved in the practical aspects of the application of research findings, as I did, control becomes an integral part of one's responsibilities. Control does not necessarily mean the killing or destruction of anything. There once was a philosophy—and still is to some extent—that if anything gets in one's way that might be termed a "pest," kill it. Gradually, we are learning that we, as one small organism on this planet, would do much better if, rather than killing things, we found a way to co-exist with them so far as possible.

I once believed that if we were going to keep our forests green and free of insect devastation, applied control should be undertaken at the earliest possible moment upon the first indication of an increase in the population of a potentially dangerous insect. I was convinced I was right in this approach. Now I know I was wrong. Experience has repeatedly demonstrated to me the fallacy of this philosophy. On the surface, such a programme sounds efficient and effective. In actual fact, it would entail the unnecessary spraying almost yearly of timber for the destruction of insect populations, which, if left alone, would be handled effectively by nature. Worse still is the danger that the application of broad spectrum chemical insecticides might intensify the problem rather than diminish it. Nature is a good manager of her own problems, and is better able in many instances to produce a solution than is man, with all his technology.

The adverse effects of chemical insecticides have encouraged the development and use of newer biological control agents. Such materials, effective only on a given target insect and totally harmless to others, as well as to all other forms of life, offer a means of safe and sensible control in the very early stages of an infestation without the dangers inherent in the application of broad spectrum chemical poisons. (See Chapter 28)

When one considers applied control of a forest insect population, one must recognize a few inherent fundamentals. First, a so-called pest of native origin, (as distinguished from a foreign-introduced organism), has existed probably for as long as the forest itself, and, in this sense, is as much a part of the forest as the trees.

Secondly, it is generally recognized that it is impossible to *totally* eliminate an insect population by any man-made means once it has become established, assuming that its environment is favourable to its existence. This applies to both native insects and foreign introductions. Their populations may be reduced to negligible proportions from time to time, but a nucleus always remains.

Living within the coniferous forest is a vast complexity of organisms, both plant and animal, co-existing in normally balanced populations. Periodically, however, there is a population explosion of one or more organisms. In the case of an insect, the results are usually severe defoliation, growth loss, dead tops and, at times, the killing of the entire tree. Their outburst in numbers accompanied by their inevitable and equally dramatic decline, is a phenomenon little understood. It is usually a complex situation involving many factors.

On the other hand natural control may be simple and direct, appearing unexpectedly to bring a sudden conclusion to an alarmingly abnormal population.

This was graphically displayed in 1961 by an unusual outbreak of a needle-feeding insect on hemlock on the coast of British Columbia near Kitimat. This defoliator, the saddle-back looper (*Ectropis crepuscularia*) spends the summer feeding on the foliage of the hemlock as a caterpillar or larva, during which time it completes its growth. With the approach of autumn, the larva drops to the forest floor where it enters the litter, or surface layer, and changes to its pupal, resting state and thus passes the winter. With the coming of spring, the adult (moth) emerges, flies to its host tree and there deposits its eggs, beginning the cycle again.

In this instance, the devastation created by the summer feeding of the larvae convinced us of the need for control the following year, provided the outbreak persisted. A survey of the overwintering population made in the autumn confirmed our fears of a very serious situation for the coming year.

The following spring, a second survey was made to determine

how the overwintering population had survived. To our utter amazement, we found the forest floor throughout the entire area of infestation torn up as though a herd of pigs had rooted through the litter. Not a sign of the overwintering population was to be found. Something had spent the winter feeding on those hibernating insects. It might have been a bird, fox, marten or other animal. Whatever its identity, it was indeed a friend of the forest.

A similar event occurred on Vancouver Island in 1970 in the vicinity of Port Alice. This involved a totally different insect but with the similarity that it, too, passed the winter in the litter of the forest floor.

We knew from previous work that a surviving nucleus of one pair per square foot of forest floor would result in a very serious increase the following year, sufficient to be a dangerous threat to the forest. A widespread survey in the fall indicated to us what we might expect in the coming year. The overwintering population numbered as high as 150 individuals per square foot. With the degree of damage already inflicted on the timber, some form of control seemed essential.

A survival survey made the following spring produced another shock. The litter was riddled with a network of small tunnels made by rodents during the winter months in search of their winter food supply—the hibernating insects. As in the previous instance, nothing remained and the insect outbreak was terminated. The specific identity of the predator remains unknown.

Although these are but two examples of natural control of an unbalanced population, observed more or less by chance, such phenomena unquestionably occur regularly in the life of a forest, albeit in a much less dramatic fashion. It is a part of the never-ending life and death struggle within nature from which results a steady density or balance of populations. In this instance, no doubt, the rodent population had been on the increase concurrently with the insect build-up for several years. With an ever-increasing food supply, rodent numbers multiplied. By eating the insect, their abundant food supply disappeared and in the process, the rodents automatically reduced their own population level. Their numbers reduced to a minimum density, the stage was set for the cycle to repeat itself.

It was a fortuitous circumstance that this event was observed coincident with rather strong recommendations that a programme should be initiated toward the destruction of the mouse and rodent population in the forest, on the grounds that they were destroyers of forest seed and hence a hindrance to regeneration, particularly where artificial seeding was being undertaken. The programme never materialized. Whether it is mouse and rodent control or the deliberate destruction of any other animal, there is one common denominator in all such programmes—the advocates are invariably only partly knowledgeable of what they are doing. We tend to blunder ahead in almost total ignorance of the very complex inter-relationships with which we are dealing. We are generally oblivious to the potential hazards surrounding us and unappreciative of the fact that within our midst are friendly organisms keeping in check what might otherwise be extremely destructive forces. While we may think we are generally omniscient about the animal life about us, we actually know very little. No one can spend his life in the field of forest biology without encountering totally unexpected, unknown or unrecorded phenomena which have existed since time immemorial.

Take, for example, a small black weevil which appeared for the first time, to our knowledge, as a great destroyer of young fir seedlings in newly-planted reforested plantations. Never before had we encountered any serious insect attack on young growth but suddenly young fir plantations were being badly decimated by a weevil which fed on the tender bark of the stems, causing the young seedlings to die. In some places, the damage was so severe much of the plantation had to be replanted. Damage was recorded in varying degrees over much of the British Columbia coast, including the Queen Charlottes and Vancouver Island.

There is an interesting thing about this insect: it cannot fly. It is strictly a pedestrian. Once, in its evolution, it did have wings and could fly like most other insects. For some reason or other, it lost this ability. Today, it looks like a plastic model with wings folded over its body, fused and immovable. Since it is distributed over the Pacific coast from Alaska to California, including the Queen Charlottes, Vancouver Island and many others, one might speculate that it reached its widespread distribution at a time when the islands and the mainland were one land mass, or that it was a great flier in the days when its wings were functional. In any case, it is not a newcomer. It has been well-established over these lands for many thousands of years. Despite its antiquity, it was generally unknown to biologists and to us as a forest pest. It announced its presence through its feeding on the bark of the newly-planted forest seedlings. Studies were undertaken to determine the reason for the "new" and unexpected injury. In due time, the story came to light.

This weevil normally lives underground at the base of trees and stumps. It comes to the surface to feed. What its preferred food is, we do not know, other than it feeds on brush or weeds growing on the forest floor.

In the course of logging, trees are felled and yarded out, resulting in a great accumulation of brush, slash and other debris which is usually burned, leaving the ground clear for the planting crews. The aim is to get the logged area restocked with seedlings before a new crop of brush can become established to compete with the growth of the planted seedlings.

The weevil, being unfamiliar with man's logging practices, comes to the surface from his home among the roots of the stumps, in search of his customary food. To his amazement, however, he finds nothing but a blackened, burned-over desert— everything he feeds on is gone. In its place, he finds young, succulent fir seedlings. Whether or not he likes fir bark as a food is beside the point. Either he eats what is put before him or he starves. Strangely enough, it was found the weevil would much prefer a piece of raw potato to a fir seedling.

So we had unwittingly created another insect problem, one better cured by controlling the cause rather than the insect.

CHAPTER TWENTY-EIGHT

Since the complete elimination of an insect population is generally considered impossible, and since even the best degree of control through the use of poisonous insecticides in the forest falls considerably short of 100 percent, it stands to reason that chemical control can do no more than reduce the population to a point where the host tree can survive, pending the appearance of some kind of natural control. This is where the situation becomes very complex.

First, the reduced numbers of insects resulting from the application of insecticide allows the tree to recover to a point where it replaces the damaged foliage through new bud and twig development in the spring. The flushing of new growth means a plentious food supply for the residual insect population not eliminated by the insecticide. Secondly, the reduced population which results from the applied insecticide eliminates any inter-competition for food among the remaining insects.

Hence, while the applied control destroys a large percentage of the insect population, it also removes one of nature's controls—starvation. In the absence of such natural controls as parasites, predators and disease, (which are usually reduced through the use of insecticide), the insect will probably rebuild its numbers to outbreak proportions in the course of three or four years, and the forest will then be threatened by a repeat of the same situation. It seems possible, therefore, that applied control may perpetuate an insect problem rather than eliminate it. The spruce budworm in Eastern Canada has been sprayed with insecticide for more than 30 years, yet the situation today is no better than it was when the project was initiated. The original objective was to stop the outbreak in the early years of the infestation. DDT was used. The budworm continued to spread. The strategy was then changed from an attempt to eliminate the insect to keeping alive the mature timber until either nature stepped in and controlled the budworm or until the timber was harvested. Today the objective is to keep alive not only the mature timber, but also the young growth that has come in since the control programme was started. In other words, to keep the forest alive, the spray programme must proceed indefinitely, or at least until a totally different concept of control and management of these eastern forests has been achieved.

Webb and Irving (1983) in their review of the entire eastern budworm situation state, "No spray involved person of our acquaintance, considers conventional budworm spraying alone to be anything like a satisfactory and ultimately tolerable way of managing the problem."[4] They postulate that the essence of the problem is how to maintain a healthy productive forest in the face of an insect pest threat; to control the insect damage, not the insect, to maintain a healthy forest and not necessarily destroying the pest. The development of eventual effective budworm management leans heavily on co-ordinated research in which physical sciences—engineering, physics, meteorology, etc., assume as much importance as the work of the biological scientist.

Most current calculations show approximately one hundred and fifty million acres (over 60 million hectares) of forest land infested with the spruce budworm between Manitoba and Newfoundland.

The common question asked is: what would have happened if spraying against the budworm had never been undertaken? The gross answer is a great deal of timber would have been killed; there would have been an enormous fire hazard within the dead stands accompanied by an equally large salvage problem; the whole eastern forest industry would have been very seriously affected.

On the other hand, the budworm would by now have run its course as in the past, and the eastern forests would be free of further budworm damage for many years to come. The millions spent on spraying could have been directed toward the salvaging of killed timber and the improvement and development of the eastern forests.

---

[4] Webb, F. E. and Irving, H. J., 1983. "My fir lady. The New Brunswick production with its facts and fancies." *Forest Chronicle*, Vol. 59(3), 1983, pp. 118-22.

Studies by the Canadian Forestry Service have shown that outbreaks of the eastern spruce budworm have occurred at intervals in these forests since the first decade of the eighteenth century. The budworm is an integral part of the eastern forests and evidence shows five major outbreaks in one region or another since the turn of the nineteenth century.

In his publication, *Aerial Control of Forest Insects in Canada*, Dr. M. L. Prebble states, "It is extremely unlikely that the application of conventional and biological insecticides within environmental constraints will be the answer. One reason for this viewpoint is that we have recently witnessed (late 1960's) a budworm explosion in New Brunswick in the face of fairly extensive spray operations."[5]

Concern over the New Brunswick spray programme is no longer limited to whether or not the chemicals are effective on the budworm population, but now includes its possible harmful effects on the residents of the province. Reye's syndrome, a rare and often fatal disease of children, first identified in 1963 by Australian pathologist, Dr. R. Reye, is suspected of being linked with the budworm spray formulation as used in New Brunswick. The prospect of having to continue spraying these forests indefinitely presents an extremely complex issue in scientific, social and political terms.

Although it is the eastern budworm that has created national interest through the past decade, there is also a spruce budworm problem in British Columbia. These two outbreaks are constantly being compared in budworm discussions. For several reasons, however, the two are vastly different.

In eastern Canada, the country is generally flat, not broken by high mountains and deep valleys as in British Columbia, and the forests of the east are much more uniform in composition and in age class. Furthermore, Douglas fir, one of the budworm's primary hosts in British Columbia, has a much greater ability to recover from severe defoliation than has balsam fir, one of the preferred hosts in the east.

[5] Prebble, M. L., 1976. "Spruce budworm in New Brunswick, attainment of objectives." In *Aerial Control of Forest Insects in Canada*, Information Canada, Department of Environment, Ottawa.

Furthermore, there is a marked difference in the use of forest sprays between eastern and western forests. Many millions of acres of forests in eastern Canada have been sprayed since the beginning of budworm control in 1952, and the programme is still continuing. In one year, 1973, more than 14 million acres were sprayed in New Brunswick and Quebec. A marked reduction in the acreage sprayed subsequently evolved and since 1980, the programme has been in the vicinity of four to six million acres. The insecticide is applied at a rate of 2.5 ounces per acre (210 g/Ha) in two applications at seven-day intervals. By comparison, the *total* acreage of forest land sprayed in British Columbia for *all* insect species since the first control operation in 1930, to date, totals slightly more than a quarter of a million acres. Few, even of those closely associated with control, realize this. It in no way suggests better management of the insect problems but results simply from the fact that B.C. is not plagued with such widespread and prolonged outbreaks of defoliating insects. This can be attributed largely to the topography of the land—high mountain ranges which separate one valley from another, variations in the climate and a mixture of tree species.

Despite these differences, there are, nevertheless, lessons to be gleaned from the eastern experience which parallel, to some extent, the problems in western Canada.

The spruce budworm in British Columbia is by no means a new arrival. Outbreaks date back over many years, and in this respect, it probably compares to the eastern insect. The most recent outbreak in B.C. developed in various areas over several years prior to 1977. By then, defoliation of the fir had become acute and entire mountain sides appeared to be dying. However, Douglas fir, on which the budworm was feeding, has a fantastic ability to recover even if almost totally denuded of foliage, provided the tree is not subsequently infested by secondary insects such as bark beetles and wood borers.

Defoliation was so severe in some areas that there was little food left on the trees for such a large budworm population. Threatened by starvation, the insect had to face another crisis as well. The female lays her eggs on the undersurface of the fir needle. With no needles on the tree, the female was out of luck when it came to laying eggs for the next year's crop of budworms. In other words, the budworm was in a very shaky situation despite the fact there was no evidence of such natural enemies as parasites, predators or disease which comprise natural control.

Extreme pressures were generated from various quarters for immediate control action. Others, equally zealous, were violently opposed to any chemical treatment. The B.C. Forest Service geared up for a large scale control undertaking, but at the last minute, opted out. Their decision was prompted, in part, by an aroused public whose opposition was based, to a large extent, on the budworm control experience of eastern Canada and also because further field studies by both the B.C. Forest Service and the federal government scientists did not support control, either economically or biologically.

The situation the following spring showed a general collapse of the population throughout the area of the proposed spray programme. The contention of many is that the cancellation of the applied control at a very critical time, permitted this collapse and prevented a repeat of the calamitous situation in eastern Canada. It is, of course, a very speculative conclusion, impossible to confirm or deny.

Despite the dramatic decrease in the heaviest infestations, the budworm has persisted in other areas of the province up to the present. Tree mortality has been minimal, but areas of infestation continue to expand. In the light of increased activity of other defoliating insects during these years, it might be speculated that we are simply passing through an era of insect abundance with abnormal activity of various species of defoliating insects, a phenomena that occurs periodically for reasons unknown or little understood.

When we consider current public awareness and concern about widespread use of chemical sprays, it is somewhat amusing to look back to a 1930 programme of control of the western hemlock looper on the Vancouver watershed. At that time, arsenic was the principal toxic agent used in insect pest control. A portion of the Vancouver watershed below the reservoir was dusted with calcium arsenite, applied at a rate of $3\frac{1}{3}$ pounds per acre. The only complaint voiced by citizens came from North Vancouver where some residents objected to the noise of the aircraft flying over their homes so early on Sunday morning. No objection was voiced to the dissemination of arsenic on the Vancouver watershed! Times change.

In my years of active participation in economic forest entomology, involving many chemical spray projects, not only those experimental in nature but all-out control programmes, I must honestly admit I have *never* seen applied chemical insecticides achieve the objective of "eliminating" the insect pest. True, I have witnessed enormous kills of feeding caterpillars within an area where the chemical was applied, accompanied by a corresponding decrease in the degree of defoliation to the benefit of the health of the tree, difficult as this is to measure. I have also seen pronounced and measurable reduction in the degree of attack by wood-boring insects on individually sprayed logs. But *never* have I seen an insect outbreak eliminated through the application of chemical sprays. In every case, the infestation persisted and eventually terminated in both sprayed and unsprayed regions at the same time, when nature decreed that such should occur.

CHAPTER TWENTY-NINE

The solution to our forest insect problems, if such there ever is, will *not* come from the development and use of chemical insecticides. As generally used today, broad spectrum insecticides, in which all species of insects are affected as well as some other animal life, are being regarded with increasing doubt and scepticism. Alternate materials and techniques of control are very much to the fore in current research.

Included in new developments of non-toxic control agents are: pheromones, bacterial and virus disease organisms, hormones, fatty acids and their salts (soap) as well as management of the crop, particularly as it relates to forestry.

*   *   *

A pheromone is a "substance externally secreted by an animal which will cause a specific behavioural response in a receiving individual of the same species."[6] Pheromones, in nature, are used to transmit various types of information signals. They may be trail pheromones, alarm pheromones, sex pheromones, territorial marking pheromones, and most important of all, aggregation pheromones.

Once the scientist has isolated the pheromone and determined its chemical structure, a synthetic pheromone can be produced, identical to the original, natural material. Synthetic pheromones have great potential in the field of control as alternatives to poisons, by manipulating insect behaviour to man's benefit. They may be used to disrupt normal behaviour by saturating the atmosphere thereby "confusing" the males so that they are unable to find females. Since very little mating ensues, the

---

[6] Borden, John H. and McLean, John, 1982. "Pheromone-based suppression of ambrosia beetles in industrial timber processing areas." In *Management of insect pests with semiochemicals, concept and practice*, E. R. Mirche, ed., U.S. Department of Agriculture, Gainsville, Florida, pp. 133-53.

181

numbers produced in the next generation will be extremely low. Wood-boring insects can be enticed from logs of high value into unmerchantable pheromone-baited "trap logs." They have potential for use in traps whereby insects are attracted and destroyed. As survey tools, they may be used to detect a pest in numbers far too small to be otherwise observed, and thus to predict impending outbreaks. Insects caught in pheromone-baited traps may also be used to justify applied control in sensitive areas after the population level has been proven, rather than proceeding blindly on uncertain estimates.

The use of pheromone-baited traps for the protection of decked logs in storage, was pioneered by MacMillan Bloedel in 1982 with striking results. This strategy has yet to be proved, however, as a method of control of insect populations under normal conditions within the vastness of the forest.

*　　*　　*

Another development in the search for alternatives is the use of native disease organisms.[7] Foremost in the group is the bacterium *Bacillus thuringiensis*, commonly referred to as Bt. This bacterial disease is specific to the larvae of Lepidopters (moths and butterflies). Within this group of insects, most of our destructive pests occur, and few are in any way beneficial. Being extremely host specific, Bt. is harmless to other insects and safe for fish and aquatic life, vertebrates and plants, and is consequently compatible with the environment.

In the control of a pest through the action of a disease, the insect must come in contact with sufficient numbers of the infecting agents to induce a diseased condition. Since most of these organisms have a relatively short life outside the host, the aim in a control operation is to disseminate sufficient organisms so the target insect cannot avoid contacting the disease.

Conventional chemical insecticides kill most of the insects, including parasites and predators which have been helping to hold in check the pest population. Massive rebounds of the pest

---

[7] Fast, P. G., 1973. "The use of bacterial insecticides in forest insect pest control." *Alternatives in Forest Insect Control, Symposium Proceedings O-P-2*, Great Lakes Forest Research Centre, Sault Ste. Marie, Ontario.

species to levels higher than those obtaining before the initial spray programme are commonplace.

Because it is a living organism, the production and use of Bt. is decidedly tricky, and many environmental factors influence its progress and development when applied in the field. Results cannot be pre-determined with the accuracy of a chemical formulation.

The first use of Bt., as a forest spray in Canada was on the Queen Charlotte Islands in 1960, when it was applied experimentally against the blackheaded budworm. The material was, at that time, in its early stages of development and little was known of its preparation and application. Results were not particularly satisfactory but positive enough to suggest much greater potential. Through the intervening years it has undergone a great deal of development and improvement, and is now used in varying degrees in both forest and agricultural pest control.

In June 1975, Bt. was applied against the tussock moth in British Columbia. In 1980 it was used against the spruce budworm on 57,000 acres in Quebec, 75,000 acres in Nova Scotia and in 1981 on a total of 140,000 acres.

Results in forest applications have been variable, from very effective to marginal. Its future as a forest insecticide, however, seems assured, with continuing improvements in the efficiency of the material and a drop in production costs.

A variety of *Bacillus*, (*Bacillus thuringiensis* var. *isralensis*), has been developed for use against mosquitoes and blackflies. Used in the United States, it has proven to provide microbial control without harmful effects on humans, domestic animals, wildlife, beneficial insects, fish or other aquatic life. It is of special value in environmentally sensitive regions.

*   *   *

Virus, like bacteria, constitute a control organism effective to a specific target host.[8] Once established under favourable conditions a viral organism will persist from year to year, spreading

---

[8] Cunningham, J. C., 1973. "The use of virus in the control of forest pests." *Alternatives in Forest Insect Control, Symposium Proceedings O-P-2*, Great Lakes Forest Research Centre, Sault Ste. Marie, Ontario.

183

from the point of original introduction. Where their propagation is satisfactory, they constitute a form of permanent control. The use of a viral organism as an insecticide is nothing new, for it is already there as a part of the ecosystem in which the target insect exists.

Viruses are widespread in nature and cause disease in almost all forms of life. The mention of the word "virus" arouses thoughts of germ warfare in the minds of many, creating a false alarm. However, viruses are very specific to a given host; in this case, the insect pest. For this reason, beneficial insects, other invertebrates, vertebrates and plants are not affected. In using a laboratory preparation of a virus insecticide, its propagation and manipulation is nothing more than assisting nature which, if left alone, would do the same thing only more slowly, giving the insect time to complete its destruction of the forest.

A noteworthy instance of a native virus affecting an insect population is found in the case of the hemlock looper in British Columbia. The looper had devastated large areas of prime timber on the B.C. coast in the early 1940's. Virus as an agent of control for this insect was little known or understood at that time. Not until after three years of severe defoliation by the insect did the disease appear among the larval population as a natural control. Within the space of about three weeks from its onset, the virus swept through the hordes of feeding caterpillars, causing a most spectacular collapse of the hemlock looper outbreak.

The disadvantage of depending on a *natural* occurrence of virus as a control agent is that it usually is effective only *after* two or three years of intensive defoliation of the trees. In 1975-1976, virus as a prepared spray was tested successfully against the tussock moth in central British Columbia during an active epidemic. In 1981, Dr. Roy Shepherd, of the Canadian Forestry Service, along with several others, applied a specially prepared virus spray in minimal amounts at the *beginning* of a potential outbreak of the tussock moth in the Similkameen Valley in central British Columbia before any noticeable defoliation had occurred. So successful was this work that Dr. Shepherd reported, "Our tests show that populations of different densities were decimated and by autumn, not one egg mass was to be found in the treated areas." (The tussock moth overwinters in the egg stage in masses of 200 to 300 eggs, covered with a gelatinous substance.)

Another interesting example of virus control relates to an introduced pest, the European spruce sawfly. For some years, this insect posed a real threat to the spruce forests of eastern Canada. In the course of bringing in parasites native to the sawfly from its point of origin in Europe, a virus disease of the sawfly was accidently introduced as well. Once established, this disease swept through the sawfly population. Now, although the European spruce sawfly extends from Manitoba to the Atlantic, it exists under permanent biological control, its numbers maintained at sufficiently low levels so as to be no longer an economic problem.

*　　*　　*

Synthetic hormones which interfere with the normal growth and development of the insect are also promising candidates as control agents, in lieu of toxic chemicals. Abnormalities induced through their use include the failure of larvae to mature and complete their metamorphosis, sterility of the female, failure to hibernate with the onset of winter and various other life pattern upsets which prove fatal to the insect.

*　　*　　*

Alternatives to poisonous chemicals are not restricted to biological agents. One of the most interesting and probably the oldest insecticide is one currently enjoying something of a renaissance—soap! To quote Dr. George Puritch of the Canadian Forestry Service, one of the leading researchers on the subject: "Fatty acids and their salts (soap) have been known for their insecticidal properties since ancient times and although they were commonly used for insect control in the early part of this century, they have seldom been used in a practical mode since 1940. With the growing public awareness of the hazards of the petrochemical insecticides on the environment, attention has again been focussed on these natural plant and animal products as alternative control agents."[9]

[9] Puritch, George S., 1978. "Biological effects of fatty acids salts on various forest insect pests." *Symposium on the Pharmacological Effects of Lipids*, AOCS Monograph No. 5, pp. 105-12.

185

The fatty acids and their soaps exert toxic effects by disrupting the delicate membranes of living cells, causing leakage of cellular materials, internal disorganization and inhibition of respiration. In insects, the compounds penetrate the body wall and enter the blood stream where they destroy the blood haemocytes.

Only a few of the many thousands of fatty acids and soaps are pesticidal. Some produce very abnormal growth and development of the insect rather than its immediate death. Others are toxic not only to insects but also to bacteria, yeasts, fungi, algae and moss. Recent research has shown that certain fatty acids react synergistically with various petrochemicals (e.g., methoxychlor) thereby enhancing their activities and permitting the use of lower concentrations. Most of the pesticidal fatty acids have low toxicity to various parasitoids, lending themselves to integrated pest management programmes.

The fatty acid compounds have been known to be effective when used against several of our more important forest pests. In their favour is the fact that they are natural plant and animal products and thus not alien to our ecosystem, being low in phytotoxicity and readily biodegradable. They have been used in one form or another against insects over the centuries, are readily available and can be processed from a variety of plant and animal sources. Perhaps the most important aspect of the fatty acid derivatives, however, is the wealth of information we already have regarding their metabolism in the human body. We eat large amounts of these compounds to provide energy and, interestingly, to inhibit the growth of pathogenic bacteria in our digestive tract. Thus, the fatty acid compounds are used in nature to control pests and promise an alternative to many petrochemical compounds.

*   *   *

Pest management as an alternative to pesticides includes not only the management of the pest as an organism, but also the management of the environment in which the pest lives in such a way as to be detrimental to the perpetuance of the organism, or at least to reduce its environmental impact to as low a level as possible. It aims at living with the pest, exploiting its weaknesses to man's advantage, but not necessarily at killing it.

Pest management is not simple. It is a long-range multi-disciplinary approach involving not only entomology, but many other facets of forest science, such as pathology, silviculture, tree physiology, soils, climate, as well as a clear and practical understanding of woods and mill operations. Since much of the expertise necessary for successful implementation of pest management programmes is gained through field experience, it requires persons dedicated to lifelong service in this work. As a basic requirement, a pest manager must be a professional forester with a sound training in the aforementioned disciplines.

We know for instance, that logs cut from trees felled in the autumn months are much more susceptible to ambrosia beetle attack than those from trees felled after the end of January. Losses can be reduced, therefore, by utilizing wood cut in fall ahead of winter- and spring-cut logs. We also know that even-aged forest, comprising a single species, is much more vulnerable to insect infestation than one of mixed species and ages. The spruce budworm of eastern Canada thrives in a forest with a high balsam fir content, its preferred food. New Brunswick softwood forests average about 40 percent balsam, ideal for budworm onslaught. The eventual reduction of balsam content of the forest, aimed toward protection from future outbreaks, becomes an example of applied forest management. The growing site too can have a pronounced influence on the growth and vigour of seedlings, and if unsuitable, may produce trees so poor as to be ready victims to insects and disease. Clearly pest management should not be treated as a corollary to other research projects. The reverse should apply. We know today's "man-made forests" are much more receptive to insect problems than are natural environments. "Under intensive management with shortened rotations, insect pests in young trees are increasingly important. To some extent, young forests solve problems of old forests. At the same time, they introduce and magnify other insect problems."[10]

Both the development of safer chemicals or their alternatives such as baits, pheromones, hormone sprays, bacterial or viral

---

[10] Furniss, R. L. and Carolin, V. M., 1977. "Western Forest Insects." U.S. Department of Agriculture, Forest Service, Miscellaneous Publications, No. 1339.

organisms, all of which are specific to certain insect species; and modifications in the management, production and harvesting of the crop concerned, provide potential for better, safer and more environmentally acceptable modes of control. It seems safe to predict that by the turn of the next century, the use of broad spectrum poisons for the control of insect pests, particularly those occurring in the forest, will be largely outdated, due to the advancement of our presently known acceptable alternatives. There will be, unquestionably, new and more environmentally compatible herbicides, as well as developments in the control of unwanted plant growth through biological means.

There is much inconsistency in the thinking of the general public on the matter of harmful chemicals. It is a *fact* that no chemical can be considered "safe." The degree of safety or toxicity depends on the method of use and the dosage administered.

To prove something is toxic involves only simple experimentation. To prove something is *not* toxic is to attempt to prove a negative, an impossibility.

Government restrictions, controls and registration for the use of pesticides and other products are aimed toward public safety. Nevertheless there have been many instances in the past where approved products subsequently have been withdrawn due to their deleterious effects when in general use. Regardless of reasons, such actions have seriously eroded the credibility of official government approvals.

Today, the public is becoming very concerned about the ever-increasing number of environmental pollutants, including pesticide spraying, dumping of chemical wastes and sewage into waterways, chemical spills, the pollutants emitted by proliferating industries and, most recently, the production of destructive acid rain as a result of the increasing use of coal as an energy source. What is generally not realized, however, is the degree of poisonous pollution to which we are unwittingly and helplessly exposed in everyday city livng.

One example of this type of pollution is found in a recent study of the effects on young children of environmental lead, resulting from lead contamination of soil. The study was undertaken in three centres of British Columbia: Trail, the location of a large zinc-lead smelter, Nelson, and Vancouver.

The concentration of lead in the soil, measured in parts per million was: Nelson, 192; Trail, 1320; Vancouver, 1545. Children aged one to six years were found to have twice the lead blood level as grade nine students, due, it is thought, to their closer proximity to the ground when playing. Although clinical examinations of affected children did not reveal any evidence of noticeable ill health the study does suggest that children living near highways or towns and cities with contaminated soil and dust may be absorbing much more lead than they should and it would be prudent to carry out detailed epidemiological studies.[11]

Nicotine is one of the most toxic chemicals and smoking has been proven to be a cancer-causing agent. Yet the same people who willingly expose themselves and their neighbours to this carcinogen loudly declaim the use of chemicals elsewhere, for health reasons.

The flood of patent medicines used by the public and of prescriptions given by some of the medical profession to their patients are additional examples of widespread chemical dissemination among the public.

The home gardener, lacking basic knowledge of what pesticides to use and how or when to use them, frequently uses masses of pesticides in varieties and amounts far in excess of sensible requirements.

Of all the pesticide users in British Columbia, no industry uses them with the regularity of agriculture. Apart from restrictions on the sale of the more undesirable chemicals through legislation, the farmer, particularly the fruit and vegetable grower, can decide for himself his own spray programme with the guidance of the Provincial Department of Agriculture. Without recourse to records of quantities used, some idea of the degree of over-use can be obtained by simply driving through the fruit-growing regions of the Okanagan during the season of spraying. The strong smell of pesticides drifting through the atmosphere, even far from any

[11] Schmitt, N., Philion, J. J., Larsen, A. A., Harnacek, M. and Lynch, A. L., 1979. *Canadian Medical Association Journal*, December 8, Vol. 121.

orchard, indicates something about the quantity being used and the extent to which residents are exposed. Nevertheless, it would be difficult, if not impossible, to maintain the high level of food production in agriculture at today's production costs without the benefit of chemical fertilizers and pesticides.

Data is not available to compare the use of herbicides in forestry with their use in agriculture, but there has been a pronounced reduction in the use of chemical herbicides in the forest over the past few years. Their use is not popular with many foresters and they are used only when there is no better economic solution. There has been a marked trend away from extensive foliar spraying of broad-leafed species such as alder and maple, for the value of such spraying is questionable.

Herbicides in the forest are, however, a silvicultural tool necessary to meet the objectives of full stocking of conifers in second growth stands in the shortest possible time. This does not imply their unrestricted and indiscriminate use by unqualified persons. Public concern regarding the environmental impact of the use of chemicals in the forest is appreciated by most experienced professional foresters, and it is generally recognized that the use of *any* chemicals should be considered only as a last resort, when no other means of control is feasible or practical.

No clear answers have emerged from the long and hard-fought battle over the use of chemical herbicides. One thing is certain, however—the conflict will continue until satisfactory alternatives are developed. Whereas notable strides have been made over the past 20 years in finding alternatives to chemical insecticides, the same cannot be said for herbicides. "There are alternatives, but because chemical pesticides over the past forty years have dominated agriculture and forest policy, alternatives have not received the attention they deserve."[12]

<hr>

[12] Kennedy, D., Ellis, P. and Young, C., 1982. "The Great Herbicide Debate." *Forest Talk*, Ministry of Forests, Information Services Branch, Victoria, B.C.

CHAPTER THIRTY

Decisions for or against pest control are made, as a rule, by those who have to pay the bill, be it government or industry. Decision-makers act on what they consider to be the best advice they can get, or on emotions of the moment. Many factors play a part in decision-making. Whether their concern is the forest, agriculture, wildlife, recreation or any other, those who feel their interests are threatened, particularly with financial loss, are eager to see some concrete action toward control. If, as frequently happens, they know nothing about population dynamics and cannot appreciate the very intricate and complex nature of the situation, they may become panicky and demand immediate action which may or may not be the best solution in the long run. Frequently, the same philosophy is applied to pest control as to fire control, and *this does not work*.

The layman must depend on the best advice he can get from those he considers experts unless he has had some past experience with the same problem. There are so many varieties of "experts" and advisors from whom to solicit advice that it becomes almost as difficult to choose the right one as for a dog to select a suitable lamp post!

First, there is the scientist who knows the pest, whether plant or animal. He knows its past and present history, its potential as a problem either present or future, and the need to institute positive action. His advice is based on knowledge gained through research and experience. His recommendations and judgment are generally sound, provided he is free of prejudice and bias.

Secondly, there is the similarly qualified scientist whose thinking is clouded by being too close to his specialty. He is blind to other aspects of the issue and may thus arrive at a very biased and prejudiced decision. The chemical control specialist may see nothing but a chemical as a solution; the ecologist looks for a solution through his own specialized field. Then there is the

scientist who ceases to think as a scientist and chooses instead to
follow the political route. Still others have a job to justify, a
department or working group to promote, a product to sell or, in
other words, a living to make through the promotion of control
programmes.

Some are hemmed in by partisan motivations and are forced
through outside pressures to support a proposal because they
have no alternative, even though inwardly they do not believe in it
and would prefer to oppose it.

Occasionally there arises the matter of a large appropriation
of money assigned to a particular project, procured through great
effort and at times through the sacrifice of other worthwhile
programmes. The return of the money or the disposal of large
quantities of unused chemical, if purchased in advance, constitutes
real embarrassment. The easiest way out of this dilemma is to go
ahead with the control programme as originally planned.

Finally, we have well-meaning and concerned consumers, vital-
ly interested in the environment and all that goes with it. These
people speak with good sense, convincing arguments and authority. They are an essential group in our society and serve the vital
function of maintaining a balance of design and action. The
disturbing thing is that, along with these well-meaning citizens,
invariably trail the extremists who lack knowledge, experience
and common sense. Their reasoning is strictly emotional and, in
becoming an irritant to all around them, they harm the very cause
they are attempting to promote. The extremists do inestimable
damage to the true conservationists by eroding the credibility of
those genuinely interested in finding practical solutions.

The layman decision-maker, as well as the thinking private
citizen, is in a quandary. Confronted with advice and recommendations from all sides, he may, in the end, take action strongly
influenced by personal emotions.

An example of such confused thinking is illustrated by the case
of the gypsy moth infestation in Vancouver in 1978. The appearance of this insect was considered by the Federal Department of
Agriculture to be a new introduction to the province. Inasmuch as
the infestation was restricted to a relatively small zone in the
Kitsilano District of Vancouver, it was hoped immediate action

might effect control or perhaps even eradication. An insecticidal spraying programme was proposed.

The programme laboured under a cumbersome organization composed of officials from the Department of Health and the City of Vancouver, representatives of the community, environmental groups and three levels of government! Judging by the daily press, one was led to wonder how much the majority of these people knew about the biology of the insect or the proposed insecticide. The programme was approved, postponed, cancelled, approved again and in the end reduced to only a fraction of its original design. Whether the actions taken were correct or not is beside the point. It provides an example of important decisions being made by unqualified people who were influenced mostly by emotional and political considerations.

Perhaps some of the fault rests with those who proposed the programme in the first place. Possibly the nature of the programme was never adequately explained to the public. It was initially suggested the spraying would be done over the residential region by helicopter. Such an application, with its inherent widespread drift, aroused the residents' ire and provoked heated protests by the Greenpeace organization. Perhaps there was insufficient information given with respect to alternative insecticides and methods. Perhaps the public was taken too much for granted. In any case, the whole programme was a mess.

The gypsy moth story did have a happy ending. The next year, the moth lived up to its gyspy name and quietly disappeared. Occasional specimens which have since been taken, suggest its permanent establishment in the province, but do not necessarily indicate a future problem.

There will always be a need for pest control of one kind or another. No blanket rule can govern decision-making; few precedents can be trusted and there is no book of instructions to guide our actions. Consideration must be given to such variables as aesthetics, environment, economics and other matters of concern to both the industry and the public.

In the case of a young growing forest, planned on a 75-year cutting cycle, the loss of eight to ten acres growth due to partial defoliation by insect attack becomes a matter of serious economic consideration, even if the tree recovers its normal health and even if it occurs only once during the 75-year rotation.

In a mature forest, on the other hand, the evaluation of partial defoliation through insect attack is quite a different thing. In this case, all the wood the tree will produce is already there. Even if the terminal part of the tree is killed (most defoliating insects feed most heavily on the upper portion of the tree), there is not necessarily any economic loss.

Factors of importance in this instance relate mostly to the salvage of injured trees, since trees so weakened become particularly susceptible to attack by secondary insects such as bark beetles and wood borers. Attack by these constitutes a death blow to the tree. Death is followed by the rotting of the sap wood, checking and splitting of the trunk as drying progresses and a steady increase in trunk breakage when felled. Time is of the essence in any such salvage programme, and this becomes increasingly acute in the damp climate of the Pacific coast. Experience has shown that about five years from the date of death of the tree is the maximum period for successful salvage. Although the heart wood of Douglas fir endures for many years, breakage becomes an increasingly important factor as drying progresses.

In our national and provincial parks, where timber is valued for reasons other than commercial ones, the significance of insect injury is an entirely different matter.

In a park or ecological reserve, a dead tree or snag is not necessarily a bad thing; indeed, much can be said in its defence. It forms a fascinating study centre of life, unlike anything to be found in green timber. It is a fundamental part of a biologically balanced forest. It is essential for such cavity-nesting birds as woodpeckers, chickadees, nuthatches and others which are among our most important predators of leaf-feeding bark and wood-infesting insects. It is the home of countless insects which infest rotting wood and bark, and becomes a source of food for woodpecking birds throughout the winter months as they tear away its surface in search of grubs on which to feed. It serves as a resonating surface for the pileated woodpecker as he drums away hoping to attract a passing female. It provides a roosting place for hundreds of birds throughout the summer months and a congregating place as flocks assemble in preparation for their migration south. Aesthetically, too, much can be said for the dead snag or tree. Its stark branches silhouette dramatically against the crimson sky of a setting sun. Finally, it is a site of interest as new flora develop around its base as the soil nutrition balance alters around its rotting roots.

There are bad things, too, about a dead tree or snag. In a fire it can become a flaming torch, throwing sparks and burning particles over considerable distances as the wind whips up the fire and spreads the flames. In wind, snow and rain it becomes a distinct hazard for breaking branches and broken tops falling from a great height can seriously injure or impale a person standing below.

In spite of it all, the presence and encouragement of bird life in a well managed forest is recognized as fundamental. The preservation of selected old snags is essential for the maintenance of a population of the many species of cavity-dwelling birds.

CHAPTER THIRTY-ONE

To cope with the ever present problems of pesticide use and other pollutants in British Columbia, a Pesticide Act was enacted in 1977 and amended in 1978. So far as this Act affects the forest, the regulations encompass all land used for forestry, transportation and public utilities, whether privately owned, Crown owned or Crown Granted. The regulations do not apply to farmers or small land owners. The objective of the legislation is to exercise control over the larger and more extensive areas where the use of pesticides can have significant environmental effects.

In 1981, the Environment Management Act was proclaimed. This Act does not impinge on the authority of any of the other Acts, but serves as an umbrella to provide consistency, consolidation and co-ordination to the Pollution Control Act, Water Act, Wildlife Act and Fisheries Act.

The control and administration of chemical programmes is under the Pesticide Control Act. Under this Act, it is required that the proponent of any control programme using insecticides, herbicides or fungicides must apply for and receive a permit and issue a notice of intent. The public can, in turn, appeal against such permit by notification to the Environmental Appeal Board, Victoria, stating the grounds for their appeal and paying a fee of $25.00. Such a hearing will be held in public before the Environmental Appeal Board.

The Environmental Appeal Board hears appeals against orders made under the Pesticide Control Act, the Pollution Control Act and the Water Act. The Fisheries Act and the Wildlife Act appeal procedures are not presently provided for under this Appeal Board. Like all appointed boards and commissions its value rests largely with the persons of whom it is composed, their objectivity and lack of bias in their decisions and their freedom from political interference.

The British Columbia Pesticide Act and the Environmental

Management Act are important forward steps in handling our pesticide problems and serve as safety valves for some of our more concerned citizens. There is, nevertheless, a desperate need for better understanding by an informed public on *all* matters of resource management.

A truly informed public will become a reality only when resource management is taught in the schools. It should be a compulsory course, following a carefully prepared text which clearly sets forth the relationship of the three primary resources: air, water and soil, with all other components of the environment.

The course I propose would include a complete definition and understanding of the environment, ecosystems, resource management and conservation, based on local examples so far as possible. It should be written on a strictly scientific level and include all our natural resources, favouring no one resource over another. The course should be objectively presented to enable the student and future citizens to learn how to judge whether or not a given project is really rewarding in the long run.

Such a concept demands freedom of information, much of it gained through government and other research, so that actual examples can be cited. Valuable information is too frequently buried and forgotten in old records or locked away in government or industrial files, its release forbidden for political, economic or public relations reasons. Secrecy of important records must end if our citizens, and particularly our students, are to obtain a clear understanding of their environment. Such a course of study would also open up communications among all users, resource managers, environmentalists and the general public.

Utilization of an area's resources for short term gains, ignoring inter-relationships, always results in serious problems for future resource management; soil is lost, water polluted, and growing sites debilitated. We pass on to our children and their descendents the results of faulty management, measured not only in dollars but also in ruined resources.

Public participation in matters of resource management is relatively new. There was a time not so long ago when forests were regarded only as a source of timber. Apart from those engaged in its harvesting, few others were particularly interested. There were no regulations. There was always plenty more timber on the other side of the mountain! The entire picture has now changed. There is no more on the other side of the mountain; in fact, there is scarcely enough to go around. An era of intensive forest management is dawning; an era when growing and management for maximum capacity per acre for a given soil productivity will be concentrated in those areas already known to be ecologically ideal for forest growth. Through modern forest technology, growth can be almost doubled compared to that which would occur on the same site under unmanaged natural conditions.

The importance of such technology in the administration of our forests becomes increasingly evident in the light of the multiple uses to which our land base must be assigned: residential, energy and transportation, water sheds, agriculture and ranching, mining, recreation, fish protection, soil protection, ecology and wild life protection, to mention only some.

Maximum use of our forest lands is no longer related only to the growing of merchantable trees. Timber production forests have one kind of value; recreational forests another; primitive and wilderness forests yet another; rural, environmental and plantation forests have their own special values and so on. These various concepts of forest use carry with them changes in values and justification for special treatments and protection plans to assure their maximum use and value.

Unfortunately, some people in positions of authority and influence who are able to effect decisions on these important issues, see values only as measured by the yardstick of their own design. Through their influence, laws can be changed and regulations bent to accommodate their own special interests.

The never-ending demand for more and more power, accompanied by the ever-increasing flooding of valuable valley bottom land for hydro development and water storage (some of it exclusively for the U.S.A.), the clearing of thousands of acres of forest land for more and more power line extensions, some parallelling similar forest clearings for highway construction, the ever-present pressure by industry for the exploitation and industrial development of our parks and wilderness sanctuaries are but a few of the things about which informed citizens are becoming very much disturbed. It is estimated that during the next 20 years, a minimum of 20 million hectares (approximately 50 million acres) of productive forest land in B.C. will be alienated from forest production to other uses.

On top of this, there is now occurring a "buy-up" of forest land for other uses, as the government of British Columbia has recently initiated a sale of crown lands.

In a brief to the government of British Columbia by the Association of British Columbia Professional Foresters, endorsed by such other organizations as the Federation of Naturalists, the B.C. Wildlife Federation, Canadian Institute of Forestry, Truck Loggers' Association, Federation of Agriculture, Cattlemen's Association plus several other forest industrial associations and unions, it was recognized that the maintenance of multiple use values cannot be guaranteed if our forest land is to be subjected to continued wholesale alienation of various kinds. Our growing population and the rising expectations of our citizens underline the need for better appreciation by citizens in general as well as by the government, of the desperate urgency for the preservation of our forest land base and for long-term resource management.

Strong pressures are already being exerted on both provincial and federal governments for the release of public park lands for industrial development. It is my belief that without strong and very active public opposition, our park lands will be slowly eroded away and future generations will find themselves committed by politicians to agreements which cannot be rescinded.

There seems little justification in mining in our designated park lands at this time when minerals could well be left for future generations to use if and when they are needed. Nor can there be any excuse for the mismanagement of lands and resources as evidenced in the hydro developments of Tweedsmuir Park. The public should not soon forget the Skagit Valley "giveaway." In this instance a permit was issued by the government of British Columbia to the City of Seattle, to allow them to flood thousands of acres of prime forest and recreational land easily accessible to residents of the City of Vancouver, in order to generate electrical power for Seattle. For more than 13 years this issue has been contested by those who value the forest and ecological qualities of this unique land. Only recently has an agreement been reached which will save the Skagit Valley for the people of B.C. But the people will have to pay for it. The price is not cheap. In order to buy Seattle out of a 99-year lease the province must provide Seattle with the power that would have been generated if flooding had been allowed, but at a price much lower than that which British Columbians pay for power. Estimated at many millions of dollars, this cost must now be passed on to future generations through increased electrical bills or through taxation. It was editorialized in the *Vancouver Sun* newspaper: "Pray that the governments have learned from this." Sounds like wishful thinking to me.

There is more than one can see, touch or smell in an undisturbed wilderness. To those who have learned to appreciate it, it offers communion with nature, the source of all life on this planet. It is a heritage for all of us to enjoy and protect.

I am enough of a realist to know that much of the wild country I have known and enjoyed cannot remain undisturbed indefinitely. Some of it has already fallen to the extension of highways, industrialization and real estate development. I would, nevertheless, like to know that in our parks, both provincial and federal, there will always be found places of undisturbed solitude and beauty such as those I have known and loved. And I would like to know that they will still be there for my children's children, for their children and for generations to come. This can only be if these natural resources are handled by people with imagination and an appreciation for values other than monetary ones. Only a determined, concerned and *informed* public can effect the preservation and perpetuation of these irreplaceable resources.

*　　*　　*

As I look back over more than 50 years of active participation in the field of forestry, I glimpse something of the great change in the profession. The whole field of forestry has become increasingly scientific; as a result there has evolved a demand for professional foresters. It is increasingly challenging to keep abreast of the times. In 1983, the Association of B.C. Professional Foresters, of which I am a charter member, numbers more than 1,450 members. Forestry today not only includes the growing, protection and utilization of the forest crop but also is closely related, in theory and practice, to the integrated use of forest land for wildlife, range land for cattle, recreation, water management and all those policies which promote the greatest social and economic returns to society. As an integral part of these principles comes the urgent need for more and more research into all aspects of such inter-related concerns.

As I look back, the thought that plagues me is: where have all the years gone? Only now do I feel equipped and experience-wise enough to live a truly productive life. If only I could begin all over again!

We are endowed with limited talent and ability and we strive through life to carve a career within the confines of these restrictions. Whether or not I have used what abilities I have to their maximum, is a matter of opinion. Of one thing I am very sure: I have enjoyed my life and the land about me. When I attempt to balance on one side my achievements and contributions to my profession and society, and on the other side the benefits I have received, I find the scale grossly weighted in my favour.

I never possessed any degree of wealth, nor was I ever destitute. Money never played a major role in my life, not that I don't want it or enjoy having some. I like it and can use it just as much as anyone else. Fortunately for me, Vi has an extremely good business head. More important to me than great wealth is freedom—freedom to organize my life as I wish and to do what interests me most. That to me is fundamental.

Today I live in so-called retirement on 35 acres of waterfront property acquired when we first came to Vancouver Island before land values commenced their rocket spiral. From the land and sea we obtain our year-round requirements of beef, poultry, fish and vegetables, and from the woods, all the fuel we need, as well as a degree of independence and freedom not easily found in these days.

Throughout my life, my family has always been extremely important to me and I was singularly fortunate in having lived in a time and manner whereby I was able to weave my professional work into the fabric of family, a family who have always had a fundamental interest in my vocation and in participating whenever and wherever they could.

During my daughter's school years, I took her alone each year for a week or more into some remote wilderness region where we camped and fished. I wanted her to grow up with a knowledge and appreciation of nature and to understand the meaning of wilderness. Perhaps it was this, along with an inherited spirit of adventure, that prompted her to venture forth alone at the age of 20 with her meagre savings, for a year's journey around the world. By freighter and hitch-hiking, she toured the Orient, the Far East, and went via the Suez to Britain to visit our ancestral home in Scotland. Hitch-hiking over continental Europe, she eventually arrived home in Nanaimo, 12 months later, rich in memories of new-found friends in faraway places and with a knowledge of foreign lands and customs few are privileged to gain.

Presently, she and her husband, Peter Grover, a principal in one of Nanaimo's schools, and their two daughters, Heather and Robyn live in our original house to which they have made some additions. Vi and I reside in a new cottage on the same property but separated from them by enough distance to assure mutual privacy. Our grandchildren will complete their schooling and will doubtless carry through their lives many happy childhood memories of summers spent on their grandparents' farm at Cedar-By-The-Sea.

Only in retrospect do certain events stand out as something special in one's life. At the time of its happening, an event is often not seen or appreciated as a rare occasion. Now, years later, such incidents are recalled and relived, triggered by a familiar sound, a smell, a taste or touch...those golden moments of the past remain alive and forever green in memory.

I still enjoy and love the land about me as I sit with my family on the patio, overlooking the pastoral landscape and the sea. The mid-summer breeze rippling through the quivering grass from the meadow below carries with it the fragrance of new-mown hay and the synchronous chirping of crickets. And all the while, our cattle are straining through the fence as they rip branches and the fruit from the apple trees; the crows in the cherry tree fight noisily over the last of the cherries, while the deer graze peacefully on the vegetables in our garden, a vivid demonstration of nature's concept of the multiple use of our resources.